Other Books in McGraw-Hill's Complete Construction Series

Bianchina ■ *Room Additions*
Carrow ■ *Energy Systems Handbook*
Powers ■ *Heating Handbook*
Vizi ■ *Forced Hot Air Furnaces: Troubleshooting and Repair*
Woodson ■ *Radiant Heating Systems: Retrofit and Installation*

Dodge Cost Guides ++ Series

All from McGraw-Hill and Marshall & Swift

Unit Cost Book
Repair & Remodel Cost Book
Electrical Cost Book

HOME AUTOMATION AND WIRING

James Gerhart

McGraw-Hill

New York San Francisco Washington, D.C. Auckland Bogotá
Caracas Lisbon London Madrid Mexico City Milan
Montreal New Delhi San Juan Singapore
Sydney Tokyo Toronto

Library of Congress Cataloging-in-Publication Data

Gerhart, James.
Home automation and wiring / James Gerhart.
p. cm.
Includes index.
ISBN 0-07-024674-2
1. Dwellings—Automation. 2. Electric wiring, Interior.
3. Dwellings—Electric equipment. I. Title.
TH4812.G47 1999
696—dc21 98-10827
CIP

McGraw-Hill

A Division of The ***McGraw-Hill*** *Companies*

1 2 3 4 5 6 7 8 9 0 DOC/DOC 9 0 4 3 2 1 0 9

ISBN 0-07-024674-2

The sponsoring editor for this book was Zoe G. Foundotos, the editing supervisor was David E. Fogarty, and the production supervisor was Sherri Souffrance. It was set in Melior per the CMS design by Michele Zito and Paul Scozzari of McGraw-Hill's Professional Book Group Hightstown composition unit.

Printed and bound by R. R. Donnelley & Sons Company.

This book is printed on recycled, acid-free paper containing a minimum of 50% recycled de-inked fiber.

CONTENTS

Preface **xiii**

Chapter 1 **Home Automation Fundamentals** **1**

Introduction 1
Power Line Carrier Control 2
Timer Controllers 2
Computer Controllers 3
Wireless Remote Control 3
Scope of Home Automation 3
Benefits to Homeowners 5
Benefits to Builders, Installers, and Service Contractors 10
Multiunit Housing 12
The Nuts and Bolts 13
Motivation 14

Chapter 2 **Control Standards** **17**

X-10 17
What Is X-10 18
Improved Communications 21
CEBus 22
What is CEBus? *22*
Home Plug and Play 24
LonWorks 24
Interoperability 25
Operation 26
EIA Standards 27
LonWorks becomes CEMA Standard EIA-709 27
Working with LonWorks 28

Smart House 28
Total Home 31
Home Automation Control 31
Infrared and Radio Frequency 32
Dedicated Controllers 37
PC Controllers 38
Interfaces 39

Chapter 3 Command Pathways 43

Power Line Wire 44
Lightning Protection and Grounding 45
Power Line Carrier Products 47
Low-Voltage Wiring 48
Category-Verified Wiring 50
Levels and Categories 51
Video Cable (Coaxial) 53
Planning a System 54
The Head End 56
Dual Coaxial Cable Wiring Systems 57
Fiber Optics 58
Wire Routing 59
Avoiding High-Voltage AC Wiring 59
Special Considerations 60
Keeping Wires Organized 62
The Automation Closet 62
Attics, Ceilings, Outside Walls, and More 64
Empty Conduits 65
Raceways 65
Wiring Ducts 66
Fiber-Optic Enclosure Systems 66
Fishing Wires 67

Chapter 4 Automating the HVAC Systems 71

Zoned Control 71
Thermostats and Humidistats 73
Considering the Needs of the Homeowners 74
Protocol 75
Temperature Control 77

Sensors 77
Converting to Digital Form 79

Electronic Duct Dampers 82

Duct Boosters 83

Air Circulation 85
Attic Ventilator 85
Whole-House Fans 85
Windows, Skylights, and Doors 86

Main Entry Doors 90

Chapter 5 Plumbing 95

Controlling Water at the Source 96

Showers and Baths 96
Control 98
Automatic Faucets 99
Automatic Toilets 101
Temperature-Balancing 104

Water Softeners 106

Magnetic Fluid Conditioners 108

Water Heaters 110

Pools and Spas 111

Laundry Areas 112
Clothes Washers 113
Clothes Dryers 114

Chapter 6 Communications 119

Telephone Lines in the Home 119
What Extra Telephone Lines Do for Homeowners 119
Benefits of Multiple Phone Lines 120

Topology 120
Star Topology (a.k.a. Home Run) 121
Ring Topology 124
Bus Topology 124
Hybrid Topology 126

Wiring for Additional Phone Lines 127
Telephone Polarity 128
Connecting Block 129

Surface-Mounted Screw Terminals 129
Type 66 Blocks 130
Running the Lines 131
Dual Coaxial Cable Wiring Systems 132
Networks 133
Wiring for Networks 134
Ethernet 134
Fiber Optics 135
Serial interface 136
EIA RS-232 Interface 137
RS-485 Interface 140
IEEE 1394 141
Overlapping Technologies 144

Chapter 7 Entertainment 149

Home Theater 149
Acoustics 150
Floors and Ceilings 152
Walls 154
Audio 154
The Dolby System 155
Home Versions of Dolby 155
Lucasfilm THX 156
Components 157
Speaker Arrangement 160
Video Components 162
Direct-View Sets 163
Rear-Projection Sets 163
Front-Projection Two-Piece Sets 164
Flat-Panel Plasma Displays and Cathode-Ray Tubes 164
Graphics Projectors 166
Projection Screens 167
Projection Screen Gain 167
Types of Screens 168
Sizing Home Theater Screens 171
Aspect Ratios 171
Common Aspect Ratios 172

Noise Interference 173
Radio Frequency Ground Breaker 175
Video Ground Breaker 176
AC Power Isolation Transformer 176
Laserdisc 176
Digital Versatile Disc 177
Digital Satellite Systems 177
Providers 179
Installation 179
Line Doublers 181
Video Distribution 182
RF Signal Integrity: 182
RF Signal Loss and Gain 184
Interactive TV 185
Interfacing 187
Control 188

Chapter 8 Home Security 193

Security Philosophies 193
Deterrence or Prevention, and Detection 194
Lighting 194
Sensors 194
Other Deterrents 195
Closed Circuit Television (CCTV) 197
Doors and Windows 198
Window Installation 198
Exterior Door Installation 200
Garage Doors 201
Fire, Smoke, and Heat Detectors 205
Carbon Monoxide 213
Automatic dialers 216
Apprehension 219
Response Time 220

Chapter 9 Lighting 225

Lighting Systems 225

Lighting Components 226
Task Components 227
Types of Lights 229
Incandescent 229
Fluorescent 230
Tungsten-Halogen 232
Low-Voltage Tungsten-Halogen 235
Fixtures 237
Recessed Lighting 237
Track Lighting 238
Wall Sconce 238
Remote-Source Lighting 239
Working with Fiber-Optic Cable 241
Exterior Lighting 242
General Installation Tips 243
Lighting Control 243
Computer Controls 244
Sensors 245

Chapter 10 More Uses for Home Automation 249

Outdoor Watering Systems 249
Installing a Full-Yard Watering System 250
Riser Types 250
Wire and Splices 257
Trench Depth 258
Flushing the System 259
Installing Sprinklers 260
Automated Sprinkler Systems 260
Watering Interior Plants 262
Pools and Spas 265
Anti-icing 266
Convergence 267
Applications 269
Transmission Types 270
Satellite 270
Twisted Pair 271
Coaxial and Fiber-Optic Cable 271
Return Path 272

Remote Control and Navigation Systems 272
New Products and Developments 273
Vocal Net 273
C-Phone Home 273
Quick Silver Hydro 274
Home Vision-PC 274
Home Director 274
Leviton's MOS 274
Denon DVD-3000 275
NAD Distribution Preamp 275
Speakercraft "Wave Plane Technology" 275
Z-Man Audio Signal Enhancer 276
Solar Light at Night 276
Summary 276

Appendix A Data Communications and Cable Manufacturers 281

Appendix B Manufacturers of Security Components and Systems 287

Appendix C Conversion Tables 293

Glossary 297

Index 315

ABOUT THE AUTHOR

James Gerhart, the author of McGraw-Hill's *Everyday Math for the Building Trades*, is widely experienced in residential construction, working for many years as a project coordinator. He built his reputation through teaching subjects such as math, estimating, scheduling, blueprint reading, and surveying to construction management studies students.

PREFACE

Home automation has emerged from the realm of science fiction to the ordinary. Porch lights that automatically come on at dusk and go off at dawn are commonplace. Security systems have been a growth industry for the past several decades, and setback thermostats became a must-have item during the oil crisis of the 1970s.

During the 1990s, the home automation industry grew from infancy into young adulthood. The development of control and communications protocols allowing exciting new products to interact with each other has opened the door to real automation. Manufacturers are introducing new products and systems almost daily. As homeowners are made aware of the convenience, comfort, safety, and status these products provide, the demand for home automation skyrockets. This demand will fuel annual revenue growth of $1.6 billion between now and the year 2002, effectively doubling the market.

The market is expected to become a giant starting in 2002 with projected revenues increasing from 3.2 billion dollars in 2002 to 10.5 billion dollars in 2008. With this type of growth, contractors, builders, and installers will be expected to be well versed in the ins and outs of home automation. Prewiring homes for computer networks, home entertainment, and multilevel communication will become an expected stage of home construction. A portion of this expanding profit center will be found in the retrofit segment of the market. Owners of existing homes are just as anxious as buyers of new homes to experience the benefits of home automation.

Home Automation and Wiring prepares the reader to take part in this growth industry while claiming a fair share of the expected $10.5 billion revenues. The book takes a systems approach to identifying and explaining the features and benefits of the various home automation technologies, as well as the nuts and bolts of design and installation.

Chapter 1 provides the reader with an introductory overview of the functions and benefits of home automation as a whole. Benefits to the homeowner are pointed out, not only for education purposes, but also for use as selling points.

Chapter 2 explains the various control standards used to convey and interpret home automation commands to individual systems and devices. Major control concepts are compared, pointing out the advantages and disadvantages of each.

Chapter 3 addresses the planning, installation, and utilization of various wiring schemes, infrared systems, and radio frequencies as command pathways.

Chapter 4 concentrates on the systems and components of home automation that regulate and control the HVAC systems and subsystems of a home.

Chapter 5 provides detail on how home automation is used to control the various elements of a home's plumbing system. Devices that regulate the supply, treatment, heating, and expulsion of water throughout the house are discussed.

Chapter 6 covers the designing and installation of effective internal and external communications systems and devices to meet the homeowner's current and future needs. Home office requirements, computer networks, and telecommunications are some of the sections covered in this chapter.

Chapter 7 details the planning, installation, and components of home entertainment. This chapter covers the design and construction of a home theater as well as the integration of home automation with entertainment systems for distribution throughout the house.

Chapter 8 focuses on utilizing the various design concepts and elements of home automation to provide safety and security. The chapter details different approaches to security as well as the benefits of integrating a security system with the whole-house automation controller.

Chapter 9 deals with the many lighting requirements in and around the house. An overview of lighting concepts and design variations are provided with details utilizing home automation to control the various lighting needs of the homeowners.

Chapter 10 covers some of the major developments adapting home automation to a variety of labor-saving devices. An overview of these technologies and their installations prepares the reader for discussions with the homeowner to determine what systems may be desired.

Appendix A is a partial list of companies manufacturing cables and wires used in home automation installations.

Appendix B is a listing of companies manufacturing security devices and controllers.

Appendix C provides some frequently used conversion tables.

James Gerhart

CHAPTER 1

Home Automation Fundamentals

Introduction

Manufacturers of labor-saving appliances have been promising homeowners an automated "home of the future" since the World's Fair days of the 1930s. Recent technology has advanced to the point where the home of the future has become a viable option for today's homeowners.

Home automation, also referred to as *home control, smart home, smart house,* or *intelligent home,* is actually a collection of devices, systems, and subsystems which have the ability to interact with one another or function independently. This allows homeowners to control almost any appliance or system in the home individually or collectively, by using automatic scheduling or by making impromptu changes.

Today more and more homes are being automated as people recognize the benefits that can be derived from automation. People enjoy automation for a variety of reasons. Some people recognize that home automation will help them cut energy costs. Others want the type of security that only home automation can offer. Still others like the control it provides while they are away from the house or vacation home.

Security systems, for example, have become a consideration for every home and a driving force behind home automation; today, a variety of security systems are available to meet most needs. Lights can be programmed to switch on and off at scheduled times or at random times when the homeowners are away, giving the home the look of being actively occupied. For convenience, lights in distant rooms or outdoors can be controlled from a central location, or the same lights can be controlled from multiple locations.

Various components that provide sophisticated home automation link advanced electronic switching technology to existing 120-volt (120-V) residential wiring for use in any residence. These components can automate security lighting and integrate it with an existing security system. Individual units are controlled by using unique coded signals.

Power Line Carrier Control

Often *power line carrier* (*PLC*) controllers transmit the unique codes over the alternating-current (ac) wiring network of the house directly to the PLC receivers. PLC components are easy to install and operate. No special wiring or other modifications are needed. These PLC components are used to switch lights and appliances from across the room, from another room, or from any convenient location in the house. The touch of a finger on a touch pad can initiate the switching sequence, or the switching can take place automatically, in response to a switching program built into a home automation controller.

Timer Controllers

Most of us are familiar with a variety of programmable timers. Timers have been on the market for decades and are useful for turning lights and small appliances on and off at preset times of the day or night. More advanced timer control units have far greater capabilities ranging from basic home control to home automation. All are easy to program and usually provide standard functions such as

- All lights on
- Panic-button switching
- Manual remote control
- Random on/off security switching

All have battery backup to maintain their programming in case of a power failure.

Computer Controllers

Today, all home automation protocols (discussed in Chap. 2) are capable of utilizing an interface with personal computers (PCs). Computers used as home automation controllers provide enormous flexibility to home automation. Homeowners can create schedules for the automation systems and "download" them automatically at predetermined times. Lights can be programmed to specific values at certain times throughout the day.

Beyond simple action-reaction and timed control of devices, computer control integrates scheduling and advanced control of X-10, infrared, and hardwired input/output devices. Intuitive event management software makes programming easy. Many advanced control features can be customized to suit particular residential applications.

Wireless Remote Control

Several radio remote control devices are available, including multidevice key fobs, handheld remotes, and wall switches. These remote controls allow homeowners to operate lights and appliances from their cars, backyards, or anywhere throughout the home. Using radio waves transmitted through walls and ceilings from anywhere inside or outside the home, these remote controls transmit command signals to a plug-in transceiver. The transceiver receives the signal and transmits it over the home's ac wiring network to the appropriate modules.

Scope of Home Automation

A home's heating and air conditioning can also be automatically controlled for greater comfort and economy through the use of home automation. Frequently, a whirlpool tub and/or a hot tub and a home office complete with fax machines, computers, and multiple phone lines are standard requirements for new construction. Kitchens are outfitted with microwave ovens, convection ovens, dishwashers, and coffee makers, among other convenient, timesaving, and effort-saving appliances. Task lighting is often used throughout the house in conjunction with accent lighting to provide sufficient illumination for reading, working, entertaining, and relaxing. Going hand in hand with the lighting programs are automation controls which will open and close designated drapes in the home, turn on the hot tub, or water the grass. All these chores can be programmed on a set schedule or accomplished with the push of a button.

Home automation provides the means to control all the technology just mentioned and more. Controls are accessed with the touch of a button or remotely from telephones within the home or anywhere in the world. These systems are custom-designed to meet a variety of individual needs including convenience, energy savings, safety, and security. A simple telephone call will allow the homeowner to enter numeric codes through the number pad on the phone, gaining access to the control program. The key is to enable various networks within the home to interact with one another to some degree. Currently there are three levels of interaction: home automation, systems integration, and intelligent home.

- *Home automation* is designed to turn a subsystem, or individual appliance, on or off according to a programmed time schedule. However, in this scheme, each device or subsystem is dealt with independently, with no two devices having a relationship with each other.
- *Systems integration* is designed to have multiple subsystems integrated into one controller. The downside to this system is that each subsystem must still function only in the way in which the manufacturer intended it to operate. Basically, it is just remote, extending it to different locations.
- In an *intelligent home* the manufactured product can be customized to meet the needs of the homeowner. The electronic architect in conjunction with the homeowner will write specific instructions to modify the product's use. Thus, the system becomes a manager rather than just a remote controller. Intelligent home systems depend on two-way communication and status feedback between all the subsystems for accurate performance.

To take advantage of the latest technology, homeowners are expecting more from builders than ever before. Builders, in order to partake of this lucrative new market, need to educate themselves and train or hire individuals as custom installers. Communication between the builder and/or the custom installer and homeowners is the key to success. Quality communications will help the builder and installer understand the needs of those who will live in the home. Communications may not be possible, as in the case of speculation homes prewired by a custom installer working for a builder. Generic prewiring may work out fine if the builder and installer accurately anticipate the needs of potential homeowners. But the best results come when the concerned parties speak on a regular basis to ensure understanding. The builder or installer designing the system needs to understand lifestyles. Do the homeowners require a home office? How about a home theater? What does the person expect the home theater to do? Other areas relate to how the homes are marketed. Are new residents likely to be retired people or families with school-age children? How many computers will be in the home? How many of those computers will require access to the Internet? The custom installer can also be thought of as an electronic architect, establishing a rapport with homeowners, discussing their needs, exploring options, and anticipating future needs. Table 1.1 can be used as a template during discussions with homeowners to determine their interests and needs.

Benefits to Homeowners

Homeowners are becoming more and more aware of the benefits of home automation. These benefits generate demand while supplying reasons (along with the profit) for builders to include home automation in all new construction while offering retrofitting for existing homes. Some of the benefits, which are immediately realized by the homeowner, are as follows:

1. *Energy savings:* Energy is used only where and when it is needed. Remote and timed control of heating, air conditioning, lighting, and appliances eliminates wasted energy. True-zoned systems and controls for air conditioning and heating permit homeowners to be comfortable in the zones they inhabit while not spending money to heat or cool spaces not in use. Adjusting the hot water heater to shut down during hours when no one is at home or while everyone is asleep will reduce utility costs considerably.

TABLE 1.1 Interests and Needs Checklist for Planning Home Automation

☐	Install home communications structured wiring to provide access to your telephone, computer LAN, audio, video, TV, and digital satellite system services from communications outlets located in several rooms.
☐	Use voice announcement to advise your visitors at home entrance that you are not at home and to invite them to leave spoken messages in the home audio and/or video recording system.
☐	View your personal computer files on unused TV channels from any home-selected TV set.
☐	View your video sources such as camera, VCR, laserdisc, or digital satellite system on unused TV channels from any home-selected TV set.
☐	Listen to the identity of phone callers on your home speakers before you answer the phone.
☐	Use a remote infrared eye to control audio and video equipment located in other rooms.
☐	Install *zoned* home speakers to allow different audio sources to be heard in different rooms.
☐	Control your selection of home lights and/or appliances from your personal computer, TV set, handheld remote, any worldwide touch phone, in-house keypad, or own voice commands.
☐	Select your home lights and/or appliances to be activated or deactivated in response to your specified programmed events.
☐	Select some of your house and/or yard lights to flash in response to specified events.
☐	Create your own ambience, lighting scenes, and activate them in response to your specified events.
☐	Save your money and energy by using thermostat setback anytime the house is unoccupied and/or on defined schedules.
☐	Save energy by turning on your water heater during off-peak hours. Save your money and energy by installing a *zoned* home HVAC system, allowing discrete room temperature control.
☐	Activate your garden irrigation water valves in response to programmed schedules and outdoor humidity conditions.

- ☐ Answer your doorbell from any house telephone and/or "hands free" by using your house intercom system.
- ☐ Have your doorbell activation forwarded to your office, cellular, or any worldwide telephone, and communicate from remote locations with any visitor at the entrance door.
- ☐ Display the phone caller ID on any home-selected TV sets and/or broadcast it over the house speaker system.
- ☐ Mute and/or pause your home audio/video system and stop your vacuum system when the phone and/or doorbell rings.
- ☐ Use your standard house telephone as an intercom.
- ☐ Allow only incoming calls on your list of authorized callers to ring your phone.
- ☐ Adjust your home temperature remotely from any worldwide telephone.
- ☐ Display your camera images of selected rooms on selected TV sets.
- ☐ Display camera images of people at the home entrance on selected TV sets when motion is detected at the door and/or the doorbell is activated, and/or record on VCR tape when the security system is armed.
- ☐ Have camera images of any vehicle entry into the driveway presented on selected TV sets and/or recorded on VCR tape when the security system is armed.
- ☐ Broadcast a voice announcement that a vehicle has entered the driveway; and/or that people are at the home entrance over the house speaker system; and/or record on audio tape when the security system is armed.
- ☐ Have the disarming of your security system cause a voice announcement over the house speaker that a security VCR or audio recording was made during your absence.
- ☐ Start a security VCR recording when motion in the yard or near the swimming pool is detected.
- ☐ Broadcast a voice announcement over the house speaker that motion in the yard or near the swimming pool is detected.
- ☐ Broadcast a voice announcement or warning to persons in the yard or near the pool.
- ☐ Cause selected lights to flash inside the house when motion is detected in the yard or near the swimming pool.

TABLE 1.1 Interests and Needs Checklist for Planning Home Automation (*Continued*)

☐	Select some home lights to turn on and off on a random schedule, making the home look lived in when unoccupied.
☐	Stop your HVAC blower in the event of fire.
☐	Turn off automatically the main house water valve in the event of a water leak and/or when the home is unoccupied.
☐	Have the carbon monoxide, gas, and/or smoke detectors activate the alarm system.
☐	Have your security system alert others on your phone call list that an emergency (such as intrusion, fire, water leak, abnormal home temperatures) has occurred.
☐	Use a handheld panic button to alert others on your phone call list that an emergency situation has occurred.

2. *Convenience:* Sound can be monitored in any area of the home or office from a remote location. Lights can be turned on or off throughout the house from a single location. Temperatures inside the residence or vacation home can be adjusted before people arrive. From one remote control in a car the homeowner can control all the garage doors and gates on the property as well as interior and exterior lights.
3. *Security:* The old-fashioned peephole in the door is rapidly being replaced by closed-circuit television (CCTV) systems which tie into the home automation program. A small, inexpensive camera discreetly placed and aimed at the door can be connected to the home automation system. This arrangement allows visitors to be viewed from any television set in the home. Using the same idea, cameras can be set up to monitor specific rooms in the house, such as an infant's bedroom. Security mode can be used to manipulate lights; heating, ventilation, and air conditioning (HVAC); sprinkler systems; televisions; and radios to make the house look and sound occupied while the homeowner is away. Garage and entry doors can be automatically checked to make sure they are shut and locked. The entire alarm system or just selected zones in or around the house can be armed and disarmed.

4. *Time and effort savings:* Control all the lights (inside and outside) from your bed at night. Program the stereo or television to automatically lower the volume when the telephone or door is answered. Preprogram lighting for various functions, such as parties, daytime, housework, sleep, or theater viewing.
5. *Comfort:* Adjust pools, hot tubs, hot water heaters, air filters, humidifiers, heating and/or air conditioning, electric blankets, and bathroom heaters all with one smart interface.
6. *Accessibility:* Unique voice- and/or switch-activated multimedia computers are being designed specifically to assist physically challenged people. For this segment of the population, these systems are not just a matter of convenience or a sales gimmick. Necessary tools, which return some of an individual's independence, these systems are being welcomed into the lives of people who are largely dependent on others to perform even the simplest tasks. These customized systems perform environmental control. From simple voice commands or touch pad switches, these systems control electric appliances such as lights, televisions, and adapted interior and exterior doors and garage doors. Plumbing fixtures such as toilets and tubs can be operated with water adjustments as illustrated in Chap. 5. Infrared appliances such as a videocassette recorder (VCR) or television can be controlled while the system raises or lowers thermostat temperatures. These systems can be used to operate normal computer functions such as word processing, faxing, and surfing the Internet. These systems can even answer and dial telephones, which allows the user to telecommute from home to a job otherwise out of reach.

Currently two products are being combined to provide extended services to homeowners. The first product is basically a *voice browser* providing voice-synthesized information from any Web site through a touch-tone telephone. The second component promises to integrate voice mail, e-mail, faxes, address books, and calendars through a natural language voice-user interface. This form of product convergence will definitely be useful when information is needed and no screen or keyboard is available, but imagine the benefits to the vision-impaired people in the population.

Integrating voice and data networks is no longer limited to innovative solutions. Networking companies along with telephone equipment providers are developing a *multiservice central office switch*. This switch, when installed in phone company central offices, facilitates new telephone-related services, such as

- Voice over IP: using the Internet as a long-distance carrier
- Internet call waiting: announcing over the Web that a voice call is coming in on the phone line being used to surf the Internet
- Internet call completion: delivering voice calls over the user's Internet connection, rather than requiring disconnection from an Internet session
- E-mail waiting: using an indicator or a "stuttering dial tone" on the phone when incoming e-mail has arrived in the Internet in-box

Benefits to Builders, Installers, and Service Contractors

Throughout history, we have always welcomed advances which made life easier. During pioneer days in the United States, to have a hand-operated water pump inside the house was considered the pinnacle of luxury. Indoor plumbing, electric lights, washing machines, clothes dryers, electric irons, and dishwashers at one point were all considered miracles of modern technology. The common desire to keep up with technological advances (and keep up with one's neighbors) has helped home automation gain mass appeal. Although the primary home automation subsystems deal with security, audio/video, HVAC systems, and lighting, more systems are being developed and marketed daily.

Many professionals in the construction industry have realized the new opportunities that home automation presents. Homes that include home automation technology have greater appeal to a broader base of buyers, ensuring greater customer satisfaction as well as faster, easier sales and reduced turnaround times on speculation homes. Homes that include automation add to the total prices of the homes, which also translates into greater profits for the builders. Subcontractors who offer home automation installations as part of their services will realize an increase in business. Most subcontractors who are already running power lines and phone lines can substantially increase profits with little extra effort by learning the ins and outs of prewiring for home automation. Combining home automation wiring with the normal wiring plan saves time and effort while adding value to the home.

Benefits to builders, installers, and service contractors are numerous. A partial list follows:

1. *Satisfied customers:* Potential customers become aware of the products and the possibilities within their own homes from many sources—books, magazines, and television as well as exposure to the home automation systems of friends and neighbors. As the availability of premium entertainment through cable and wireless sources makes the home a primary venue of entertainment, the demand for home theater escalates. More people are working at home than ever before, relying on telecommunications, computers, and local-area networks (LANs), all requiring special wiring and installations. As more builders preinstall wiring for home networks during construction, it becomes easier and cheaper for a homeowner to take that first step into home automation. The builder who provides customers with access to new technology, which permits homeowners to pursue their work and leisure activities, will have word-of-mouth advertising worth a fortune.
2. *Additional profit centers:* From 1998 to the year 2002, annual revenues in the U.S. home automation market are expected to double, resulting in an increase of $1.6 billion. Growth through the year 2008 is predicted to run at 19 percent. From 2002 to 2003, experts believe the market will triple, growing from $3.2 billion in 2002 to $10.5 billion in 2008, which translates to an increase of $7.3 billion over a 5-year period.
3. *Year-round work opportunities:* Given the increased demand for home automation systems, the builders, installers, and service contractors can be busy all year. During seasonal lulls, the company's emphasis can be shifted to indoor work on new construction and retrofit jobs. Home automation can provide enough work to maintain or even increase profit levels while keeping work crews together year-round.

4. *Increased market segment:* General contractors, subcontractors, and electricians are among those who can easily increase their share of the construction market. As discussed earlier, the home automation market is expected to realize some remarkable growth in the next 5 years. How much of that $7.3 billion market could go to your company?
5. *Automatic expansion:* As with computers, stereo systems, and technology as a whole, home automation will continue to evolve. New products will continue to be introduced to the market, and of course, someone will be needed to install, update, and service these new products. As the market grows, so will your business. Many home automation modules are of the "plug and play" variety. Others are equipped with color-coded leads for fast, simple installation. Where appropriate, these modules fit easily into the same electrical boxes that hold standard switches and receptacles.

Multiunit Housing

Developers and contractors involved in, or thinking of getting involved in, the multiunit (apartment) housing market need to include home automation in the plans in order to stay competitive. In recent years, apartment and condominium communities throughout North America have been built to cater to the specific needs of two market segments. The first group is young, single (or newly married), and well employed. The second group has most of or all the same characteristics as the first group. However, the second group is defined by a desire to buy a home but an unwillingness to settle for anything less than a dream home with all the conveniences.

As stated previously, more people are working from home now than ever before, and those who do go to an office frequently bring work home to complete in the evenings or on weekends. Sophisticated communications technology, video, and on-line shopping services make life easier. ISDN, cable modems, T-1 lines, 10baseT networks, and other methods of accessing the Internet and the "information superhighway" are becoming major buying decisions for this segment of the population along with greater labor-saving devices around the home.

Developments that have the appropriate wiring already installed to enable this capability have been very quick sellouts. In addition, many contractors equip each unit with in-wall stereo speakers for whole-home background music. Many communities come prewired for home theater, with installed surround-sound speakers in two- and three-bedroom residences. By using category 5 wire for telephone and high-speed data transmission, each apartment or condominium can be preconfigured for three to four telephone numbers. In place of the standard telephone outlet, each room has a multimedia wallplate, equipped with a telephone jack, a data jack, and two video jacks supplied by high-performance RG-6 quad-shielded coaxial cable (Fig. 1.1). Living rooms are equipped with speaker connections for bedroom background speakers and living room surround-sound speakers.

The Nuts and Bolts

Each unit is normally equipped with a custom-designed structured wiring distribution panel, usually installed in a bedroom closet. The distribution panel is the gateway for all the home automation and communication services. The panel enables maintenance personnel to configure telecommunications, video, Internet, and other electronic services to the specific needs of individual tenants. As technology advances, it will permit the introduction of new services as well as future upgrades.

In addition, each unit is necessarily equipped with a whole-house surge protector. The surge protector has a light indicating that

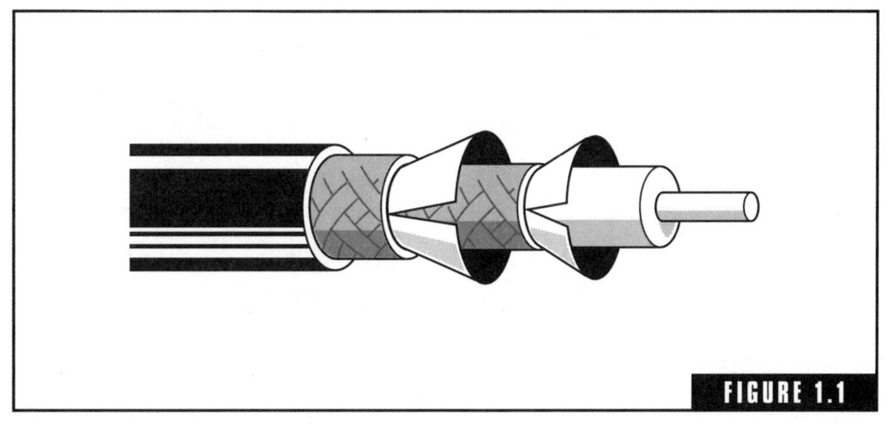

FIGURE 1.1

RG-6 quad-shielded coaxial cable.

the surge protection is active. The whole-house surge protector works with secondary surge protectors for equipment such as computers and home audio equipment to prevent damage from electrical surges. Power line filters and couplers are also generally located inside the circuit breaker panel. These devices facilitate the use of equipment using the electrical wiring in the apartments for control and communication.

Motivation

With rental fees continuing to rise, developers need to set themselves apart from the competition in order to reach full occupancy. Whether one is building single-family homes or multiunit complexes, home automation makes a residence more appealing, practical, economical, and desirable. These factors translate to higher selling (or rental) prices and a larger profit.

As this book continues, the applications and convenience of home automation will quickly demonstrate why homeowners are demanding these features in their homes.

Notes

Notes

CHAPTER 2

Control Standards

The functions of home automation systems are carried out by using one or more control standards to interpret commands and convey them to the designated device(s) or subsystem. Various manufacturers pioneered home automation products using their own proprietary control standards. Because of the proprietary nature of these standards, products made by one manufacturer were not usually compatible with those of other manufacturers. More recently, the development of nonproprietary standards, combined with participation by giants such as AMP and Lucent, has resulted in standardized components and architectures that allow easier planning, development, and installation of systems.

Several control architectures have been adapted into nonproprietary standards and are prevalent in most products. In this chapter we discuss the major concepts of the most popular and familiar names of home automation control standards on the market today.

X-10

One of the first protocols to be developed is known as *X-10.* Pico Electronics Ltd., a Scottish firm, was the developer of several chips (single-chip calculator integrated-circuit chips) for projects that were known by production codes X-1 through X-9. In 1976, X-10 resulted from Pico's experimenting with the idea of consumers controlling lights and appliances remotely without having to rewire the home.

What Is X-10?

Since the X-10 architecture controls lights and appliances through existing power line wires, the system uses the 60-hertz (60-Hz) power (or 50-Hz power in some countries) as a carrier. The X-10 technology functions by modulating a 120-kilohertz (120-kHz) burst. The presence of a burst equates to a digital 1, and the absence of a burst equates to a digital 0.

Pico Electronics also created a simple addressing protocol, which allows each unit in a common wiring environment to be identified individually. This protocol involves 16 address groups called *house codes* and 16 individual addresses called *unit codes*. Between the two, there are 256 possible unique addresses. Among the functions of the original X-10 protocol were command strings, including ON, OFF, DIM, BRIGHT, ALL LIGHTS ON, and ALL UNITS OFF. All the receiver modules within the system can "hear" these command strings, but only those set to the last address sent will respond to the command(s).

RECEIVERS

As the name implies, these modules receive the command strings over power line wiring and carry out the task indicated by the signal (i.e., turn on/off or dim in the case of lights). To install a receiver, simply connect it to the electrical wiring system. Connections can be made by either plugging the module into an electric outlet or replacing an existing switch with an X-10 wall switch. By adjusting the built-in dials controlling address code on each individual module, the receiver will understand which X-10 signal to respond to. Some of the more popular receivers are listed in Table 2.1.

Because the protocol involves 1 bit of information, transmitted on each ac cycle, the transmission rate in a 60-Hz environment is 60 bits per second (bits/s) or 50 bits/s in a 50-Hz environment. To send a signal that includes both address and function, the protocol requires 48 bits, or 0.8 s in a 60-Hz environment, and just under 1 s in a 50-Hz environment. To send a new function string to the last address requires one-half that time.

TABLE 2.1 Popular X-10 Receiving Modules

Receiver	Wiring	Actions	Applications
Appliance module	Plug in	ON/OFF	Turn electrical appliances, stereo, TV, hot water tank, hot tub ON or OFF.
Lamp module	Plug in	ON/OFF/DIM	Turn lamps ON/OFF and DIM/BRIGHT.
Wall switch	Replace existing	ON/OFF/DIM	Turn lights on the existing circuit ON/OFF and DIM/BRIGHT. Three-way switches also available.
Wall receptacle (electric outlet)	Replace existing	ON/OFF/DIM	Control whatever is plugged into the outlet ON/OFF and DIM/BRIGHT.
Universal module	Plug in	OPEN/CLOSE dry contacts	Control low-voltage systems such as sprinkler valves, drapery controls, garage door opener.
Chime module	Plug in	CHIME	Chimes when ON signal received.
Siren modules	Plug in	SIREN	Security alarm horn activated by X-10 security system.

Today, X-10 is a home automation standard as well as the name of its manufacturer. X-10 produces a line of branded products and X-10-compatible products for client companies. A wide range of products use the X-10 technology, including many types of light switches, remote controls, security systems, television and computer interfaces, and telephone responders. Some of the most commonly used X-10 controllers are listed in Table 2.2.

Historically, X-10 has not been the answer to every home automation requirement. Many X-10 products (or modules) communicate in only one direction. For example, many modules receive commands and turn themselves on or off or dim, but they do not transmit. Consequently, other modules within the network are unaware of an appliance's status. On the other side of the coin, many X-10 transmitters (controllers) send out control signals but do not receive feedback from receiver modules. The 256 addresses and six functions of the X-10 standard also create built-in limits for consumer products. Additional addresses and functions are limited and offered only on particular types of products. These issues are being addressed with the introduction of two-way lamp and appliance modules which include status

TABLE 2.2 Some of the Most Common X-10 Controllers

Controller	Wiring	Actions	Applications
Mini/maxi controller	Plug in	ON/OFF/DIM and all ON/OFF	Simple ON/OFF/DIM control of lights and appliances from fixed remote location.
Minitimer	Plug in	Manual and timed ON/OFF/DIM	Schedule ON/OFF times for various lights and appliances (security mode controls lights at random times)
Wireless remote control	Plug-in receiver	ON/OFF/DIM	Remote control sends radio signals to plugged-in receiver which triggers X-10 equipment. Range of 75 feet (ft).
Wireless wall switch	Plug-in receiver	ON/OFF/DIM	Velcro-mounted keypad sends radio signals to plugged-in receiver which triggers X-10 equipment.
Telephone responder	Plug—telephone line and electric outlet	ON/OFF/ and all ON/OFF	Dial in from anywhere in the world and use touch-tone buttons to control X-10 devices.
Computer controllers	Plug—serial port and electric outlet	Programmable events	Program sequences of events to occur in response to remote signals and times.

polling, modules offering automatic status response, and modules utilizing automatic gain control (AGC). These new modules are the first in a series, working in conjunction with the X-10 ActiveHome two-way computer interface and software.

TWO-WAY MODULES

The two-way modules are controlled by using standard X-10 commands. Once the command has been carried out, the module can then be polled for its status.

STATUS REQUEST

When a module is polled for its status, it responds with its on/off status in the same code format as that of existing specialty two-way modules, such as transceiver modules and floodlight motion sensors. Using the same code allows existing software and hardware to automatically work with these new two-way modules.

AUTOMATIC STATUS RESPONSE

Two-way modules can also be configured from the next generation of ActiveHome software to automatically transmit their status when turned on, whether by a computer interface and software or by any other X-10 controller. A change in status will also be reported when an appliance is turned on or off locally (manually at the appliance).

OUTPUT MONITORING

When a receiver is sent a command, the acknowledgment is sent not upon receipt of the command, but only after that command has been successfully carried out by the module. The receiving module also monitors the output status of the relay or switching circuit; therefore, any failure of the hardware will be transmitted back to the controller when polled.

While X-10 offers some infrared (IR) and radio-frequency (RF) controller and receiver modules, standards for media such as twisted-pair, coaxial cable, and fiber optics are not provided for within the X-10 protocol.

Improved Communications

Manufacturers, designers, and users of home automation are constantly seeking ways to lower costs and improve the functionality and flexibility of this technology. Open, interoperable systems are meeting these desires. An *interoperable system* integrates multivendor controls for HVAC, lighting, fire, safety, security, entertainment, and general convenience. Standard, interoperable control technology results in lower installation costs and introduces efficiencies and savings throughout the system life cycle.

To successfully integrate products made by different manufacturers requires the use of unified communications. An interoperable system eliminates the need for gateway options such as custom hardware and software. While an open communication technology supports a variety of new products, systems operating costs are also reduced. This new technology also provides the flexibility to take advantage of improved subsystem integration and reduced life-cycle costs by changing or modifying systems in the future.

CEBus

Once home automation products began to proliferate, the need for a home automation controller which would include functions beyond ON, OFF, DIM, BRIGHT, ALL UNITS ON, and ALL UNITS OFF was recognized by members of the Electronics Industry Association (EIA). Beginning in 1984, engineers representing companies from around the world worked together to develop a proposed standard. What they came up with was named *Consumer Electronic Bus,* or *CEBus* (pronounced "see-bus"). Products that incorporate CEBus standard technology will communicate with one another to provide consumers with enhanced control, comfort, and convenience in the home. In March 1998, the Consumer Electronics Manufacturers Association (CEMA), a sector of the EIA, published the first 12 CEBus standards in the ANSI/EIA-600 series. The EIA-600 series was designed to accommodate the following functions:

- Remote control
- Status indication
- Remote instrumentation
- Energy management
- Security enhancement
- Entertainment device coordination
- In-home distribution of audio and video

While serving the needs of existing control communication requirements, CEBus is also intended to be flexible enough to work with anticipated future requirements at a practical cost.

What Is CEBus?

CEBus is an open-architecture set of specification documents which define protocols for how to make products communicate through power line wires, low-voltage twisted-pair wires, coaxial cable, infrared signals, radio-frequency signals, and fiber-optic cables.

Since CEBus is a nonproprietary standard, anyone anywhere can get a copy of the plans and can develop products that work with the CEBus standard. CEBus works by using device addresses that are preset in the hardware at the factory; these addresses allow 4 billion variables. The standard also offers a defined language for many object-oriented controls including the commands

- Volume up
- Fast forward
- Rewind
- Pause
- Skip
- Temperature up or down by as little as 1 degree

One of the hallmarks of the CEBus standard is this Common Application Language (CAL), defined in EIA 600.81. This document presents the language and an object-oriented model of interoperability between diverse and otherwise unrelated products that constitute the home environment. A generic version of CAL is being developed so that products designed for other protocol stacks can place this protocol-independent CAL element within the application layer of their new and existing products and thereby expand the product's features and functionality.

CEBus transmissions are based upon strings or packets of data that vary in length depending upon how many data are included. The CEBus standard includes such things as spread spectrum modulations on the power line. *Spread spectrum modulations* involve starting a modulation at one frequency and altering the frequency during its cycle. The CEBus power line standard begins each burst at 100 kHz and increases linearly to 400 kHz during a 100-microsecond (100-μs) duration. A burst is referred to as the *superior* state, and the absence of burst is referred to as the *inferior* state. The digit 1 is created by either an inferior or superior state that lasts 100 μs. The digit 0 is created by an inferior or superior state that lasts 200 μs. Since both the superior and inferior states create similar digits, a pause in between is not necessary.

The rate of transmission is variable; depending upon how many characters are 1s and how many are 0s, some packets can be hundreds of bits in length. The minimum packet size is 64 bits, which at an average rate of 7500 bits/s would require approximately 0.0085 s to be transmitted by the controller and received by the target device or subsystem.

The CEBus trademark is owned by the EIA and a group known as the CIC (CEBus Industry Council), which is currently pursuing further product developments. The CIC is a nonprofit organization made up of representatives of many national and international electronics firms, including such notables as Microsoft, IBM, Compaq Computer Corp., AT&T Bell Laboratories, Honeywell, Panasonic, Sony, Thomson Consumer Electronics, Leviton, and Pacific Gas & Electric.

Home Plug and Play

The Home Plug and Play (HomePnP) Task Force is a previously private-sector initiative formed as an informal work group to address interoperability issues related to local-area networks (LANs). The Home Plug and Play Task Force recently announced the selection of the CEBus Industry Council to complete the development of the HomePnP specification and promote its adoption by manufacturers of electronic products for the home.

The specification provides clear guidelines for developing interoperable home automation and home office products while simplifying the purchase, installation, and management of home LAN products, allowing incremental installation by homeowners and professional installers. An example of networking occurs when the homeowner opens the automatic garage door. As the garage door opens, it communicates to other systems that the residents are returning home. Lights illuminating a walkway for the homeowners may come on as well as interior lights or appliances, and air conditioning or heating may be automatically adjusted for the residents' comfort. Other networking examples include doors that automatically lock at night and appliances that announce when a part is going to fail. Home computers will become multitasking LAN servers monitoring messages sent between household products and carrying out homeowners' wishes. There are endless other applications of interoperability that will be developed and marketed in the near future, demonstrating the convenience and safety of home automation.

LonWorks

LonWorks is a control network technology. The main emphasis of this technology, introduced in 1991, deals with the distributed control aspect of automation networks. LonWorks networks are becoming the dominant standard for home networking around the world. One of the key factors in making this happen is energy management. Currently, there are more than 25 utilities located throughout the United States and around the world that are pursuing projects using products based on LonWorks networks. In addition, numerous OEMs and systems integrators have announced LonWorks-based products and services for the utility industry. The projects represent a wide range of utility applications [e.g., demand-side management (DSM), meter reading, home automation, distribution automation, substation automation, and generation plant control] and address all types of utility customers (e.g., residential, commercial, and industrial). A key strength of LonWorks networks is their acceptance across all areas of a utility's control networking needs.

LonWorks provides the tools and building blocks to build, install, and maintain intelligent, interoperable nodes, subsystems, and systems. Providing an open platform, LonWorks makes it possible for the tools, modules, and integrated circuits (ICs) to be readily available from multiple suppliers, and they can be further supplemented by third parties as needed.

Interoperability

In August 1994, thirty-six companies banded together to found the LonMark Interoperability Association (LIA) to facilitate the development and implementation of open, interoperable LonWorks-based control products and systems. The membership has grown to more than 200 today and includes end users, integrators, manufacturers, and virtually every major controls company in the building industry worldwide. The LIA provides an open forum for member companies to work together on marketing and technical programs to advance the LonMark standard for open, interoperable controls. Its three major functions are

- Definition of design guidelines for interoperable devices based on LonWorks technology
- Certification of products that meet the LonMark standard for interoperability
- Promotion of LonMark products and systems as open, interoperable control solutions

The LIA develops technical design guidelines for products to ensure that they can be easily integrated into multivendor LonWorks networks using standard tools and components. These guidelines require the use of standard data objects and types along with self-documentation to support easy installation and configuration of systems using third-party tools.

Operation

In a LonWorks network, intelligent control devices, called *nodes,* communicate with one another using a common protocol. Each node contains embedded intelligence that implements the protocol to monitor and/or control functions. There may be 3 or 30,000 or more nodes in a network: sensors (temperature, pressure, passive infrared, etc.), actuators (switches, ballast controllers, circuit breakers, pool pumps, etc.), operator interfaces (displays, touch-screen terminals, human-machine interfaces, etc.), and controllers (room controllers, set-top boxes, utility gateways). Each LonWorks node includes local computational and networking resources and can also attach to local input/output (I/O) devices. The I/O resources allow it to process input data from sensors or output data to actuators. The computational resources allow it to process the data for signal conditioning or as a part of the control loop. The networking resources allow it to interact with other nodes on the network to complete the control loop. By eliminating the need for the central controller or computer—often the Achilles heel of conventional control system hierarchies—distributed control has been shown to enhance system reliability, responsiveness, and predictability while lowering installation and life-cycle costs.

Communication tasks are managed through a powerful integrated circuit that allows devices to make decisions and communicate those decisions to other intelligent devices in the network through one or more communications media. The media can consist of twisted-pair wire, power lines, coaxial cable, optical fiber, or wireless methods such as radio-frequency and infrared. The flexibility to support one or more media makes LonWorks networks an attractive solution for retrofits where existing twisted-pair, structured cabling, or power lines may be used.

To a degree, LonWorks nodes on a LonWorks network are analogous to personal computers on a data network. However, unlike data networks, which are optimized for large file transfers on high-bandwidth media, control networks send small control messages originating with sensors that contain status information instead of large files of data. Actuators on the network observe this status information and make a local decision as to what kind of action to take with the damper, valve, or other device to maintain control of the system. Devices in the network are installed in the field using off-the-shelf network management software running on Windows 95, Windows NT, or other platforms. Once devices are installed, the network manager either can remain connected to the network or can be disconnected. Since the intelligence is distributed in the field devices, no supervisory controller is required for the system to operate.

The benefits of open, interoperable, multivendor systems are

- Choice of vendors
- Use of third-party tools
- Easier integration
- Reduced installation costs
- Easier additions and changes

These benefits free end users from lengthy, costly service and upgrade agreements with a single vendor. Now end users can implement control systems using LonMark devices from multiple vendors, confident that whatever they choose can be easily integrated.

EIA Standards

The EIA's Integrated Home Systems (IHS) technical committee was created to look at alternative standards for home networking. On August 29, 1996, the EIA formed a subcommittee to consider an additional EIA standard for home networking based on LonWorks network technology.

LonWorks Becomes CEMA Standard EIA-709

In 1998 the Consumer Electronics Manufacturers Association (CEMA) published a new standard for home networking, EIA-709, based on the LonWorks control networking platform. Divided into three parts, the LonWorks-based EIA-709 standard defines a communication protocol for networked control systems in a home (EIA-709.1) and transceivers for networking consumer products over existing power lines using narrow-band signaling (EIA-709.2) and free-topology twisted-pair media (EIA-709.3).

- EIA-709.2 defines physical communication over power lines inside and outside of homes over 120-V-ac to 240-V-ac wiring. The power line channel occupies the bandwidth from 125 to 140 kHz and communicates at 5.65 kbits/s. The standard supports both two- and three-phase electrical configurations.
- EIA-709.2 calls out a narrow-band power line signaling technology that meets the regulatory requirements for North America and the European Community.
- EIA-709.3 supports free topology communications, including bus, star, loop, or any combination of topologies at 78.125 kbits/s. The network cable conforms to ANSI EIA/TIA 568A category 5 (a 24-AWG unshielded twisted-pair cable that is commonly used for structured wiring systems).

Working with LonWorks

With more than 1500 shipping products and more than 3500 control system suppliers in the development of product, there is a large and growing need for qualified integrators. To help the building industry respond to the growing demand for interoperable solutions, Echelon Corporation, a Palo Alto, California, manufacturer of hardware and software products for the development and implementation of control networks, is introducing the *Network Integrator Program.* This program provides the necessary products, tools, and training to help network integrators install, commission, and maintain LonWorks interoperable solutions.

The growing availability of products displaying the LonMark logo is making it easier for specifiers and integrators to create lower-cost, more flexible, and more efficient automated solutions today.

Smart House

In 1984 the U.S. Congress passed the Cooperative Research and Development Act, allowing collaboration between companies for purposes of research and development (R&D), in a private consortium without violating antitrust laws. The National Association of Home Builders (NAHB) in Washington, D.C., formed the Smart House Limited Partnership (L.P.). The NAHB then invited companies to consider membership in Smart House. The intent of Smart House was to allow three competing manufacturers per Smart House component or product to join the consortium.

Smart House has about 25 manufacturers who have signed formal contracts, called *research and licensing agreements,* to develop products. Another 25 companies have formed business affiliations with Smart House to develop applications. Most of the capital for Smart House was provided by builders and loan guarantees from the NAHB. In 1992, two manufacturers of electrical connectors—AMP, Inc. and Molex—teamed up to buy all the rights to the Smart House technology and to provide additional funds for Smart House, L.P. to continue system development.

An overview of this technology is significant historically because it was the first widely marketed integrated home automation system. The fundamental goal of Smart House is to provide integrated wiring for all current home services and to include provisions for home automation technology. Cost savings are promised from reduced labor to install an integrated wiring system.

Smart House is designed for specific home automation applications, such as entertainment, lighting, HVAC control, and the interface between an existing security system and a Smart House control panel. Smart House uses a proprietary system design. One of the goals of the Smart House concept was the creation of a unified wiring bundle that would substitute for the diverse collection of wires that general contractors usually have to deal with. Instead of having the many different trades responsible for the electric wiring, telephone wiring, security system wiring, cable television wiring, and so forth, a single electrician could handle all the wiring, thus lowering the total cost of the job and easing the problem of scheduling conflicts between different installers. Smart House wiring consists of three cable groups:

1. *Branch cabling:* power plus digital data. The branch cabling includes a conventional power cable and a digital data cable to minimize mutual interference and reduce costs. The digital cable consists of four pairs of twisted-pair wires.

2. *Applications cable:* digital data plus direct-current (dc) voltage for sensors.

3. *Communications cable:* video coaxial cable plus telephone wires.

SMART HOUSE COMMUNICATIONS

Digital signaling wires accommodate appliance control, status, and message data. The system is laid out with branches corresponding to the present electric branch circuits, as seen in Fig. 2.1.

A system controller is located at the hub of the system. Physically, the system controller is placed adjacent to the electric load center containing the circuit breakers. This location is called the *service center* because it also contains a noninterruptible power supply, surge suppressors, ground-fault circuit interrupters, the head end for the coaxial cable system, and the telephone gateway.

The system controller performs system management functions, handles the routing of data between appliances, and provides scheduling for selected services in the house. The use of a system controller

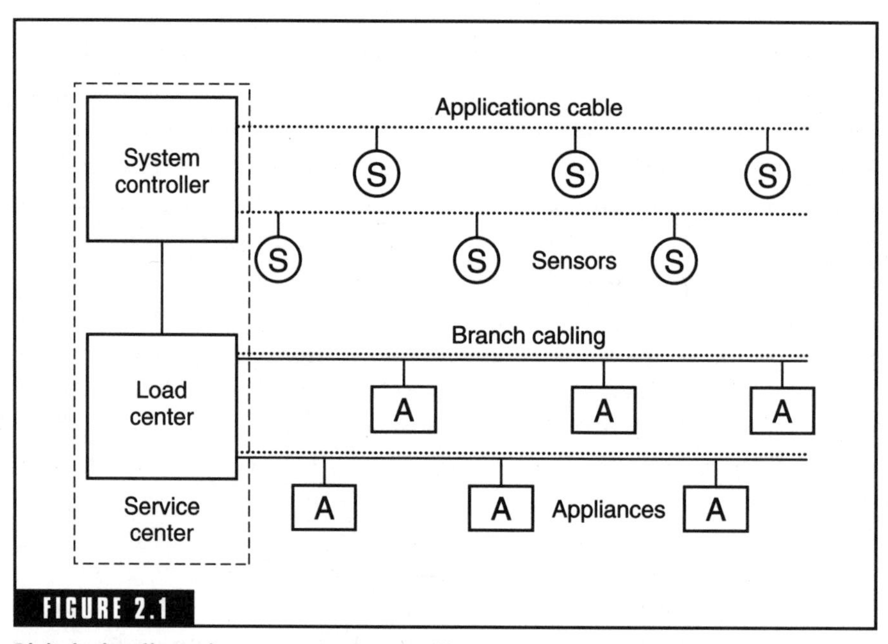

FIGURE 2.1

Digital signaling wires accommodate appliance control, status, and message data. Example of Smart House wiring layout.

for all message transfers is different from its use in most other home automation infrastructures.

Smart House is very proprietary and dealer-oriented. Installation and service to the systems require training, certification, and specialized tools and hardware. Many larger companies, such as AT&T, Sears, Honeywell, and Square D, were once involved with Smart House and have since left. Originally proposed as a standard which included centralized intelligence and new wiring and hardware for new homes, Smart House has recently begun to work with both CEBus and LonWorks in an effort to develop hardware and components intended for existing homes. It is expected that someday soon Smart House will change from the centralized-intelligence concept to a CEBus- and/or LonWorks-based network concept. When this occurs, Smart House will have changed status from a proposed standard to a brand name.

Total Home

Total Home is a basic home automation package developed and marketed by Honeywell. Known as a manufacturer of thermostats and HVAC systems, Honeywell has also been involved with building automation in the commercial sector for a number of years. Building on the automation segment of its business, Honeywell began to offer residential security systems complete with monitoring services. Using X-10 for lighting controls and its own proprietary HVAC and security systems, Honeywell developed a subgroup known as *Total Home.* The basic Total Home package includes 10 points of lighting control, 10 security sensors, and three zones of HVAC control.

Currently, Honeywell is very dealer-oriented. Consumers cannot purchase, install, or program most of the Honeywell system components. However, the company is involved with both CEBus and LonWorks, and has stated that it will adapt to either or both standards once they become widely accepted by consumers.

Table 2.3 has been included to provide comparison of the major home automation systems at a glance.

Home Automation Control

Regardless of the control standards being used within a home automation system, one or more methods of issuing and interpreting commands must be in place. These controls can be as simple as motion, light, or temperature detectors; they can also involve a variety of components including one or more computers. In this section we introduce the main concepts of some of these control technologies.

Infrared and Radio Frequency

While the four major power line standards—X-10, CEBus, LonWorks, and Smart House—support power line communications, most also support other communications media, such as infrared (IR) and radio-frequency (RF). IR and RF systems are somewhat unique in that they are "wireless" control systems. Neither of these controls is new to us. For years now, IR controllers have made life a little easier when it comes to accessing the functions of our televisions, or operating the videotape recorder. In the early years of remote-controlled televisions, RF signals were sent to the set to change channels, turn the set on or off, and eventually raise or lower volume.

INFRARED

Infrared systems send codes on a beam of infrared light through the air from the transmitter to the receiver, where the codes are interpreted and the commands are carried out. The drawback of IR is the ease with which the signal is blocked. IR signals cannot go through solid objects such as walls. Even a family member or pet standing between the transmitter and receiver will block the signal. This problem is currently being addressed with the use of IR repeaters.

INFRARED REPEATERS

In-home audio and video distribution networks allow users to view movies or television programs in rooms other than those where the original signals emanate. For example, if a VCR is playing a tape in the living room, it can be watched on the television in the bedroom. But if the user wants to pause the movie, or fast forward, or skip to the next track, it usually means going to the living room to use the infrared remote to perform the function.

CEBus and LonWorks standards promise a convenient resource for remote control. As these standards evolve and are used in new televisions, computers, receivers, and other equipment, these devices will have the ability to communicate with one another through the power

TABLE 2.3 Comparison of Major Home Automation Systems

	X-10	CEBus (EIA IS-60)	LonWorks	Smart House
Developer	X-10 (USA) Inc.	Electronics Industry Association (EIA). Further developed by *CEBus Industry Council*	Echelon Corp. testing and certification programs led by LonMark Interoperability Association	Smart House Limited Partnership (Smart House L.P.) for the National Association of Home Builders
Media supported	Power line. X-10 manufactures devices for other media, but there are no standards for them	Power line Twisted pair Coaxial cable RF IR Eventually fiber optic	Power line Twisted pair RF Third-party transceivers support others	Custom-made wiring available from three sources
Maximum data rate	60 bits/s	10 kbits/s, additional support for video, audio, and data	610 bits/s to 1.25 Mbits/s	50 kbits/s, additional support for coaxial distribution
License requirements	Proprietary, company does not license others to use it.	Public domain, does not require a license. Certification required to use CEBus logo.	License required. Certification required to use the LonMark logo.	License required
Relative cost	Low	Low to moderate	Low to moderate	Moderate to high
Target applications	Existing and new homes	Existing and new homes	Existing and new homes, commercial and industrial buildings, automation, automotives	Mostly new homes, some light commercial buildings

line wiring, coaxial cable, or any other wires to which they are both attached. Users will access remote control functions from the room they are in; the device in that room will receive that signal and transmit it to the VCR in another room. Repeaters extend remote control to other rooms, using the existing wiring without any special hardware or setup requirements.

Before, repeaters were built into televisions and other electronics; now, stand-alone repeaters are available. Tabletop models and wall-mounted repeaters can be placed to perform the same functions. Infrared repeaters commonly use low-voltage wiring, coaxial cables, or radio waves to communicate to remote rooms.

LOW-VOLTAGE INFRARED REPEATERS

The most common type of infrared repeater uses low-voltage wiring. Infrared repeaters built into the walls of a home are connected behind the wall with dedicated low-voltage wiring. In these custom installations, an infrared receiver is often added in the wall near the door and light switches, and infrared transmitters (also known as emitters) are placed on or near the entertainment equipment in remote rooms. By aiming a standard infrared remote at the receiver on the wall, the signal is conducted to the remote room and reproduced to control the devices. These installations frequently include keypads on the same wall with the receiver. This provides another convenient control source to users as they enter or leave a room.

The category 5 low-voltage wiring in a whole house wiring system can support an infrared repeater as well. An infrared receiver is simply plugged into a wall plate in one room, and an infrared transmitter is plugged into the wall plate of another room. At the outlets and the hub, the wires are attached to an RJ-45 jack. As illustrated in Fig. 2.2, both receivers and transmitters lead into an infrared hub. For the IR bus, three wires are needed: power, ground, and the IR signal. Figure 2.3 shows a typical wire selection for this purpose.

Even though transmitters and receivers normally require only 5 V to run, selecting a 12-V power supply and equipping each transmitter and receiver with a 5-V regulator will “future-proof” the installation against new products that might require more than 5 V. Another reason for using a higher-output power supply is to ensure that transmitters and receivers receive the 5 V they require. Often voltage drops

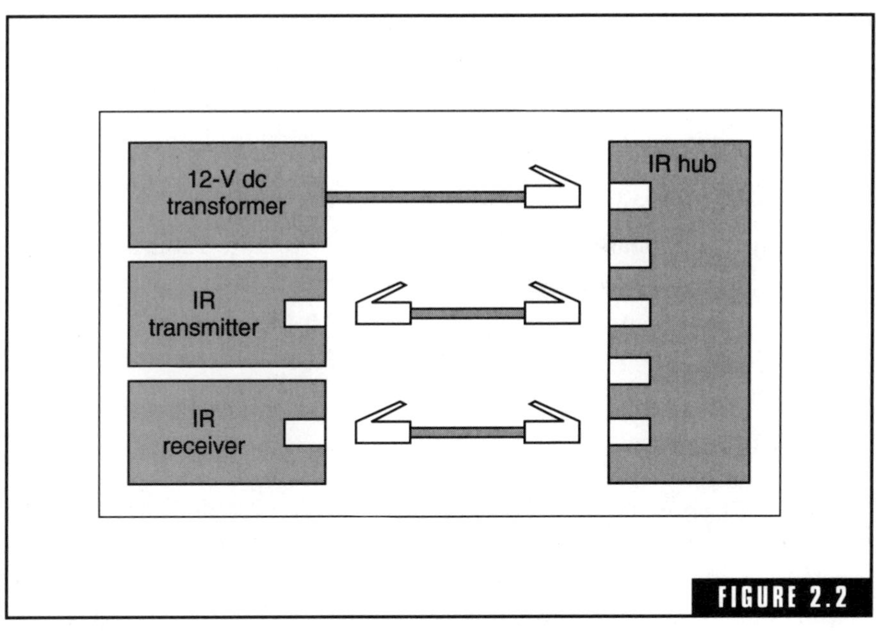

FIGURE 2.2

Both receivers and transmitters lead into an infrared hub.

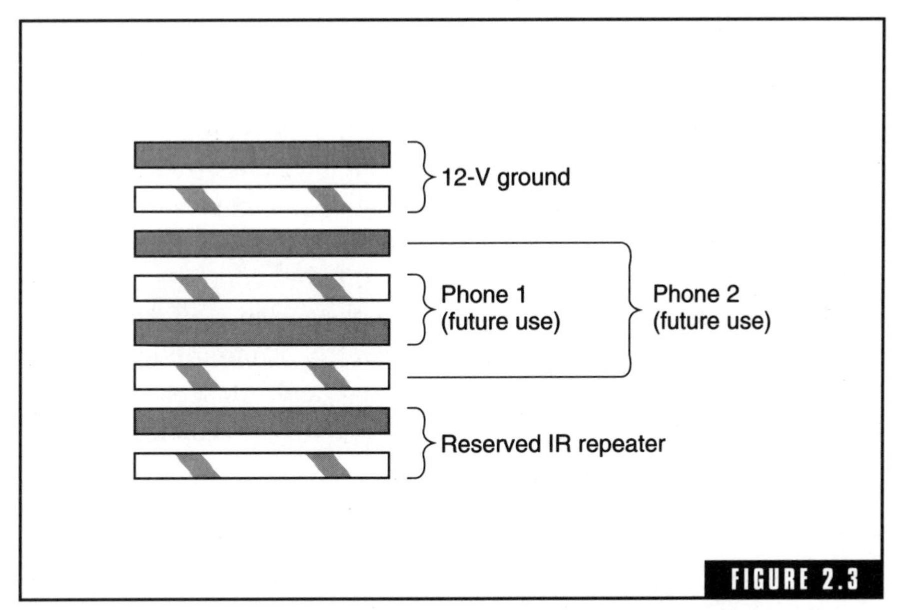

FIGURE 2.3

A typical wire selection for the IR bus. Four-pair cable signal descriptions.

through long cable lengths reduce voltage enough that the components cannot function. Generally, the power supply is a 12-V dc wall transformer with an RJ-45 plug attached. It is simply plugged into any jack of the hub.

COAXIAL INFRARED REPEATERS

Some infrared repeaters distribute signals to remote rooms using the same coaxial cable, which carries the broadcast signal. This involves infrared receivers and transmitters similar to those described earlier. This system also requires a few special splitters and combiners to add and extract the signal from the cable.

RF INFRARED REPEATERS

Infrared repeaters, which use radio frequencies to communicate, are low-cost and easy to install. Some have an infrared receiver that works with standard infrared remote controls. An infrared receiver placed on a television or tabletop includes a built-in radio transmitter that broadcasts the control signal to the emitters in remote rooms. When these emitters receive the radio signals, they reproduce the infrared signals to control devices in their rooms. Most stand-alone repeaters use this technology.

Other RF infrared repeaters are specialized RF remote controls, which can broadcast directly to the emitters in the remote rooms. This type of equipment is especially useful outdoors or in other applications in which direct sunlight exposure can interfere with the infrared signals from standard remotes.

If there are several similar entertainment devices, they may all respond to the same control signals. An infrared controller can overcome this condition. These devices allow the selection of one specific entertainment device from many and control it individually. Some infrared repeaters include built-in infrared controllers, and they are also commonly offered individually for cabinet mounting or placement on a convenient tabletop.

RF SYSTEMS

RF systems do not have the same problem penetrating solid objects as IR systems do. Radio-frequency signals can readily pass through walls, people, and doors (as long as the signal is strong enough). The strength of an RF signal depends on the power of the transmitter, the distance the signal needs to travel to the receiver, and the ability of the receiver to "pull in" the signal. Radio-frequency signals have been used for a long time, and they are currently the basis of television and radio broadcasting, security signaling, cordless and cellular telephones and pagers, garage door openers, and many other applications. RF signals have more bandwidth available and therefore fewer performance limitations than power line systems, as well as being truly wireless. RF signaling does have limitations, however. Privacy is a serious issue, as the signals do not stop at house boundaries. Bandwidth is a precious and regulated commodity, with the Federal Communications Commission (FCC) reluctant to give away too much at a time. Cost is still an issue, although devices are getting continually less expensive and requiring less power. RF technology is a useful alternative to wired systems, although in most cases, it will be cheaper and more reliable to install wired devices, if the wiring already exists.

Dedicated Controllers

Manufacturers of early home automation created centralized intelligence systems, which commonly involved dedicated computers as automation controllers. These devices range from about the size of a paperback book up to the size of a breaker panel. The size and intricacy of the home automation system to be controlled dictate the size of the controller, which is packed with electronic boards and wiring. These controllers are still in use today, primarily popular with users who are not yet familiar or comfortable with PCs. Dedicated automation controllers often supported only one of the major control standards. However, major manufacturers have developed or are developing interfaces to allow dedicated controllers to accommodate multiple standards. Since dedicated automation controllers are primarily focused on the needs of the home, most include a wide variety of built-in communication devices and features.

Supporting power line and/or low-voltage media, some dedicated automation controllers require a hybrid cable and other specialized types of wiring. Optional interfaces are used to add communications to the other media such as infrared, RF, coaxial cable, or fiber optics. Be sure to read all information supplied by the manufacturer for specifications of wiring and installation.

Although they are often used in place of PC controllers, PCs can program most dedicated automation controllers. This feature provides access for remote service providers to program or modify them. (This capability is also provided to PC automation controllers; however, it is the dedicated controller users who are not yet comfortable with PCs that are the most likely to utilize the programming services of a remote provider.)

PC Controllers

The use of a PC to control home automation is preferred by many homeowners. Possibly due in part to the fact that they may already have a computer, this type of control also allows homeowners to upgrade the automation controller by simply copying the appropriate files. Home automation control software can run in the foreground or background of the computer's processor. This means that users can leave the controller software active while they are using the word processor or one of the other programs on the computer.

USER-INTERFACE SOFTWARE

Some automation controller software allows either the installer or the user to program by writing lines of code in basic computer languages. Other software providers offer an icon-based *graphical-user interface* (GUI) such as Windows or Mac's use; these GUIs allow the user to program by a point-and-click method with a mouse. Some software also provides selections to the user in the form of menus, and others use icons, which can be placed on a background drawing of the floor plans or pictures of the rooms.

The most common controller functions include the following:

- *Timed events:* Automation controllers include a built-in clock and the ability to automatically activate or deactivate other devices or systems at predetermined times. Many automation controllers provide a *security* feature, which can activate or deactivate devices at random times within a predetermined period. This feature also allows the user to schedule events to occur each day, week, month, or year, or only on specific days.
- *Groups and scenes:* Automation controllers allow the users to create "scenes" which involve several devices that respond together. For example, although they may be connected to different switches and wiring, several lights can be turned on and adjusted to any particular dimming levels whenever a particular button is pressed or when a particular condition is sensed.

- *Event responses:* Automation controllers can have predetermined responses to a specific condition or group of conditions. For example, when motion is detected in a room, the automation controller can turn on one or more lights; or when motion is detected and it is light outside, the automation controller can open draperies, start the coffee brewing, and turn on the morning news on the television.
- *Support house modes:* House modes involve specific conditions under which the event responses will occur. For example, when the house mode is set to "home," the automation controller might provide a response to a motion sensor in which a light is turned on; and when the house mode is set to "away," the controller can flash lights when motion is detected. The most common house modes include "home and away," "awake and asleep," "day and night," "normal," and "vacation."
- *Monitoring events:* Automation controllers can log information about activities. Homeowners can review that log after leaving a service person in the home alone, to see what rooms where entered or what activities occurred.
- *Bridge between multiple media and standards:* Automation controllers can act as a bridge which receives signals from devices or systems on one medium (i.e., power line, low voltage, or infrared) or standard (say, X-10, CEBus, or LonWorks) and retransmits a comparable signal to devices utilizing another medium or standard.

Many automation controllers provide most of the functions described above, but not all. Some automation controllers may only support a limited number of event responses or scenes, and many others may only support one or more communication media and/or standards.

Interfaces

Personal computers offer the unique feature of easy interfacing with a variety of media and standards. Many of these communication interfaces come in the form of boards for internal use or as external devices that are plugged into a serial port in the computer. There are a variety of PC interfaces available. Some communicate to one medium (i.e., power line devices using the X-10 standard), while others communicate to a variety of media and standards (i.e., power line and low-voltage devices using the X-10 and CEBus standards).

Newer computers include an energy management or instant-on feature, which permits them to be left on all the time. This feature allows the computer to turn the monitor off and go into a "dormant" stage. The computer reactivates itself quickly when it senses a need for home controls or telephone answering and other functions. New computer hardware includes built-in home control hardware and software developed to allow the PC to communicate to power line devices through its own power cord. The communications hardware is built into the computer's internal power supply, and special hardware additions are not necessary.

The different forms and functions provided by these various standards and controllers depend on their use and placement. This area of home automation deals with command pathways and will be addressed in the next chapter.

Notes

Notes

CHAPTER 3

Command Pathways

Command pathways, as the name implies, are the routes over and through which commands, communications, and data are brought into and directed throughout the house. Some of these pathways are used within a home network to exchange data between members of the network and to communicate with networks outside the home. Various types of wire and cable are used for this purpose as well as light beams and air itself. In the past, the telephone company would make one connection to the home with a single pair of low-voltage wires. Today, many different companies offer the same service, competing with the telephone company. These competitors offer to route signals into the home through television cable, satellites, and radio waves.

Each of these formats has its advantages and disadvantages. A well-designed and installed system of pathways will allow signals to be easily distributed to all the users throughout the home. In this chapter we discuss the design, installation, and utilization of these pathways.

The need for a well-thought-out system of distribution throughout the home is a primary consideration today and will be even more important in the future. A generation ago, it was not uncommon for a home to have only two appliances that required special communications wiring. This included cable television (CATV) for a single television in one room and telephone wiring for one telephone and perhaps a couple of extension phones in the home. Today, most homes have several televisions and multiple telephones. Thirty percent of homes in the United States now house a computer, which at some point will most likely use the same communications system. Seventeen percent of U.S. households have more than one computer in the home. In addition, many homes are equipped with one or more fax machines, VCRs, answering machines, and intercoms, again using this same wiring system. For most homes, this means there is already an average of more than one communication device for each room of the house. In addition, these same wires and cables are often used for devices such as closed-circuit television (CCTV) cameras near the front door, pool, or nursery; doorbell intercoms; and touch-screen displays installed on or inside walls.

Proper planning for this type of growing use requires that a home designed to keep up with its occupants include more than one communication outlet in each room. Many architects, builders, and contractors are now including communication outlets on each wall of each room; this provides greater overall convenience along with greater flexibility for interior decorating.

Power Line Wire

Residential electric wiring has changed very little in the past decade and is still closely regulated by the National Electrical Code (NEC). In most cases, an electrical contractor will install electric wiring. However, there are a number of items that the architect, builder, contractor, remodeler, or custom home electronics installer should discuss with the homeowner and the electrician to prepare the job properly for today and for future system upgrades. For example, most electronic equipment is very sensitive to ground loops and ground-induced noise. A proper earth ground at the building service entrance is the first step in avoiding such problems. In many cases, a proper earth ground is provided by a connection to the steel rebar in the building's foundation. If the ground at the site is dry, a solid copper ground rod should be driven below the water table near the service entrance. In extreme cases, an active ground should be provided. This consists of a hollow rod filled with conductive mineral salts that dissolve in the soil to improve conductivity. All outside service grounds must be solidly connected to this ground point, including power, telephone, and cable television.

Lightning Protection and Grounding

For lightning protection, the common lightning rod installation (called the *Faraday cage method*) consists of multiple terminals fixed at all salient points on the roof, bonded together with copper conductor that runs down the house and into the earth ground. All antenna masts should be grounded as well. An alternative to having lightning rods sticking up every 5 ft along the ridgeline of the roof is the Franklin cone. The Franklin cone, seen in Fig. 3.1, requires only one 10-ft spike with an 8-inch (8-in) diameter pan-shaped "preventer" attached to the top. This installation is sufficient to protect a 9000 square foot (ft^2) home. Mounting it beside the chimney can hide most of the shaft.

A Faraday cage or the Franklin cone can protect the building but not the individual electronics inside. The power spike can still get channeled onto the wiring and into the house, wiping out electronics. A No. 6 AWG or larger grounding conductor should be run from the main ground to any electronics subsystem locations. Ground connection points from the telephone system controller, security alarm panel, audio equipment, and other electronics gear should be connected to this ground bus.

To ensure that all electronic equipment grounds are at the exact same electrical potential and help avoid differences in grounds that cause ground loops, all electronics and computer equipment should have a separate isolated electrical subpanel with isolated ground receptacles provided at all locations remote from the main. This is particularly true for big-screen televisions, video projectors, and remotely located audio amplifiers. Components that cannot have *equal-potential* grounds should have signals that are isolated from one another. This can be expensive and difficult to achieve, and it's much easier to prevent the problems in the first place.

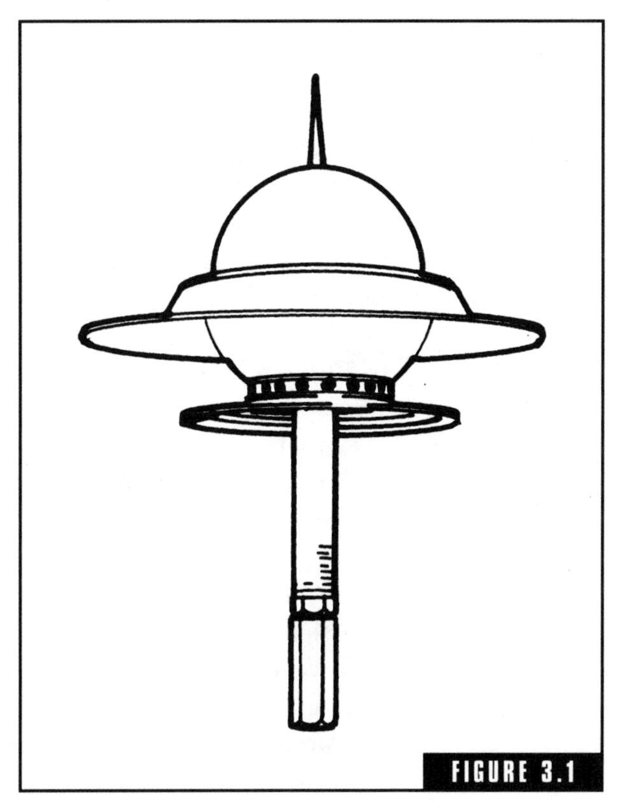

FIGURE 3.1

The Franklin cone.

The use of power line carrier devices (such as X-10 or Leviton controls) in the installation requires additional care in designing the electrical distribution system. Although these devices can be used in three-phase systems, a single-phase system is less complex and more cost-effective. Proper signal coupling between the two "legs" of a single-phase system is the key to proper power line carrier operation.

Passive signal couplers should be installed at the service entrance and at each subpanel. These require a single-gang electrical box adjacent to the panel, wired to a pair of dedicated circuit breakers (one on each leg) and the neutral bus.

Tip: An alternative to placing the signal couplers inside the panel (a violation of the NEC) is to install a *whole-house blocker* at the service entrance. This will prevent power line devices at a neighbor's house from affecting your system, and vice versa. It also acts as a signal coupler.

For long runs or to include detached garages and out buildings, an amplified coupler may be required. Keep in mind that your bid should include an exception if amplified couplers are required; this will account for the additional expense. For power line devices to operate properly, a neutral conductor needs to be run through every device and switch box in the system. Three- and four-way switches are available in power line devices, but they can be tricky to install. Better results can often be obtained by using a regular switch or dimmer and installing a controller module at the other location. The controller modules can also control up to three additional devices without additional wiring.

Surge protectors used on electronic gear often include noise filters. Unfortunately, these filters can actually filter out control signals traveling on the power line. Other common household appliances can also interfere with power line carrier devices. Some of the most troublesome include wireless intercoms, television sets, compact fluorescent lamps, motors, and personal computers. If the devices are plugged into a household outlet, an X-10 plug-in noise filter may eliminate the interference. Compact fluorescent lamps or fixtures used in the same system as the power line carrier or infrared devices will generally cause a problem. Compact fluorescent fixtures are efficient, but generally cause a great deal of both electrical and optical interference.

Surge protection is absolutely critical in any electronics installation. Power surges or nearby lightning strikes can cause sudden or delayed failure in most electronic components. Most plug-in protectors use a combination of *metal-oxide varistors* (*MOVs*) and fuses to provide protection against catastrophic surges.

Preventing signal-level problems when installing a computer interface for power line devices can be accomplished with the use of two power strips, one with surge and noise suppression and one without. The unsuppressed power strip is plugged into a wall outlet. The X-10 noise filter is then plugged into this power strip, and the suppressed power strip is then plugged into the X-10 noise filter. Computers, monitors, and printers are plugged into the suppressed strip while power line interfaces are plugged into the unsuppressed strip. The filter will suppress noise while providing 120-kHz impedance (the power line carrier frequency) which will prevent the noise filter and the computer power supply from short-circuiting the power line data.

Several manufacturers make whole-house surge protectors that install at the service entrance across the power conductors. These devices clamp the incoming voltage to a safe level and are highly recommended. For complex or critical systems, an *uninterruptible power supply* (*UPS*) should be installed. A UPS is basically a battery pack, designed to supply enough power to allow a computer operator time to save files and properly shut down the computer. Separate UPSs can be used for other critical loads and subsystems which are sensitive to power outages. Most UPS devices will pass power line signals, but computer interfaces may need to connect to the other side.

Power Line Carrier Products

This section describes how *power line carrier* (*PLC*) products can be utilized. Whether one is building a new house, adding three-way control to an existing light, or installing lighting, for landscapes the PLC can simplify ac wiring. The savings can more than pay for the cost of the PLC devices—plus provide a flexible light-level control system ready for timers, remote control, and home automation. As seen in Fig. 3.2, a PLC-based system can control lights in conventionally wired homes. By replacing the light switches (represented by S in Fig. 3.2) with PLC-activated switches, the lights can be controlled locally, at the switches, by wired or wireless remote control, or from a home

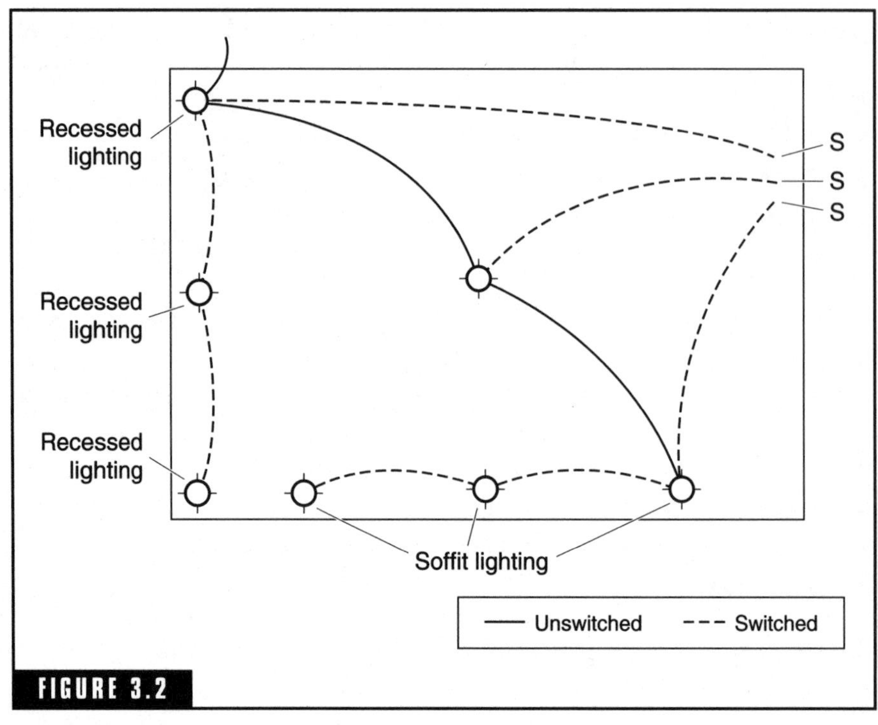

FIGURE 3.2

A PLC-based system can control lights in conventionally wired homes.

automation. Switch modules can control an incandescent load, fluorescent loads, motors, and even ceiling fans. Three- and four-way switches are also available.

Figure 3.3 demonstrates the functionality that can be achieved by using fixture-mounted PLC receivers. The T represents a PLC transmitter. A four-button transmitter can turn on, turn off, brighten, and dim three PLC addresses. Each address can be configured to affect multiple devices. The R represents PLC receivers wired directly into the light fixtures. These modules do the actual control of the lights. Notice the simplified wiring and the single-gang transmitter (instead of the three-gang switch panel shown previously). Additional transmitters placed at desired locations add three-way control.

Low-Voltage Wiring

In recent years wiring standards for telecommunication devices in the home have become much more sophisticated. The simple quad-wire (four-wire, nontwisted telephone station wire) that was standard for so many years is no longer acceptable for modern residential systems. Regulations prior to 1980 made it illegal for anyone other than telephone company employees to connect phone wiring or phone jacks. A major change was made when the Federal Communications Commission (FCC) issued wiring docket 88-57, allowing non-telephone company installers to connect to telephone company wiring systems. These days, the phone company is normally not involved in wiring work within a residence. The phone company's wiring responsibility usually stops at a point close to where the telephone line enters the home, referred to as the *demarcation point.*

There are two wiring schemes commonly used in residences: series and star. The series method, often called *daisy chaining,* was the standard wiring method used by telephone companies prior to 1980. This method simply consists of stringing a telephone wire from jack to jack. When the telephone lines needed to be reconfigured to accommodate a change in the number of lines serving a residence, daisy chaining presented a problem.

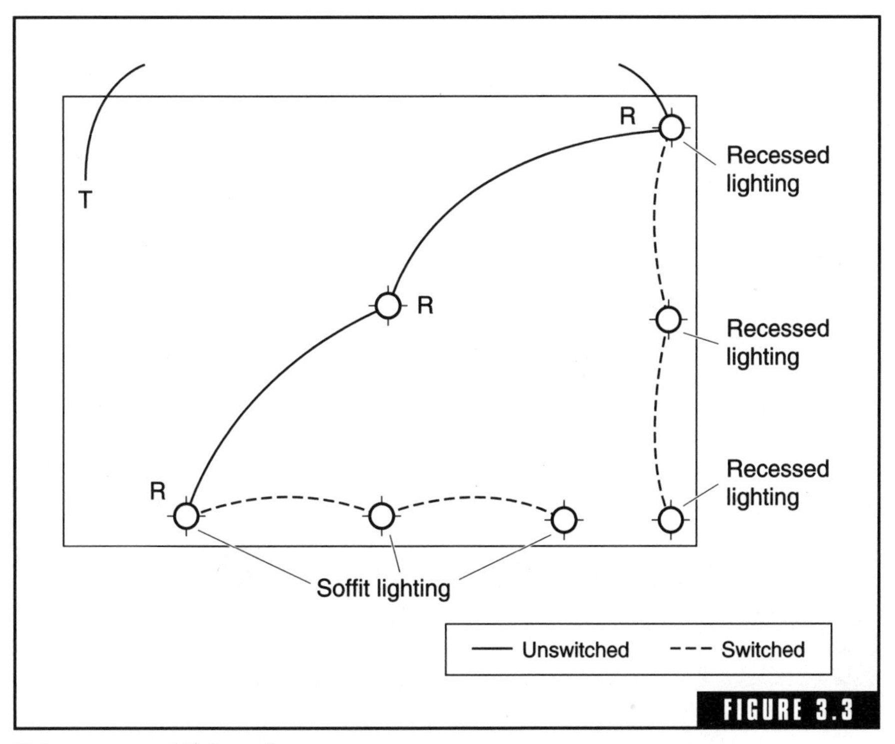

Fixture-mounted PLC receivers.

The star, or home run, method involves running phone lines from a central point located at or near the point of demarcation to each jack or room in the house. By running lines to each room from a common connecting point, it is easy to configure each line to serve any desired point in the house as more lines are added. A separate line added for the computer can be connected to a line serving the home office, and a new line to serve the children can be connected to jacks in their rooms. Reconfiguring the connections at the central distribution point is all that is needed. Figure 3.4 is an example of the home run residential wiring method. It is a modified star configuration. A cable containing four twisted pairs is fed from each room to a common connection panel, usually placed in a utility closet. Jacks in the room can be connected to any of the four pairs.

Category-Verified Wiring

In the days when telephone use was less demanding, simple quad wire was usually used to service just one line in the house with maybe an extension or two. With the proliferation of modems, answering machines,

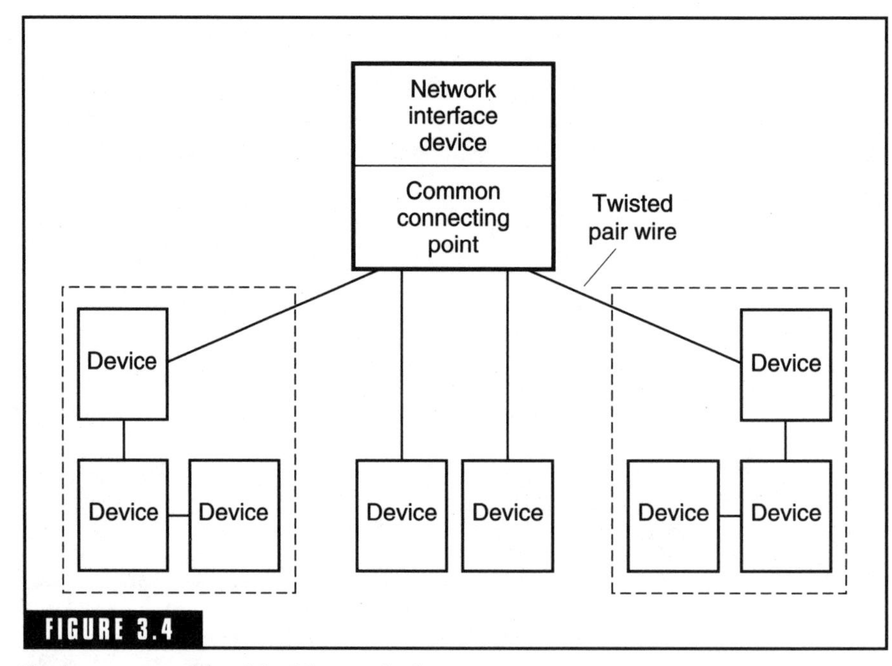

FIGURE 3.4

The home run residential wiring method.

residential PABX systems, and residential computer networks, quad wire no longer has the circuit capacity, noise immunity, or bandwidth to satisfy today's and tomorrow's residential communication needs.

To provide standards for telecommunication wiring, the Telecommunications Industry Association (TIA) established several specifications. For residential installations, *TIA-570 Residential and Light Commercial Wiring Standard* is the document for reference. Under these standards, the stipulated wire for residential use is category 3, four-pair telecommunications wire. However, for current and future home automation purposes, category 5 wire is now the mandated choice.

Levels and Categories

Levels and categories have been established to describe the performance capabilities of different telecommunication wiring methods. Before TIA established a rating system, equipment distributors developed a system of levels. The only levels that are still in use are the following:

- Level 1: Plain old telephone service (POTS)
- Level 2: IBM type 3 cabling system

More recently, a system of categories has been established by TIA:

- Category 3 = 16 megahertz (MHz) [10 megabits per second (Mbits/s)] 100-ohm (100-Ω) unshielded twisted-pair wire
- Category 4 = 20 MHz (16 Mbits/s) 100-Ω unshielded twisted-pair wire
- Category 5 = 100 MHz (100 Mbits/s) 100-Ω unshielded twisted-pair wire

Category 3 has been the standard for residential installations in the past. It is satisfactory for voice transmissions, but that is about the limit. Category 5 is fast becoming the new standard for residential use due to its greater capacity and shielding properties. Integrated Services Digital Network* (ISDN) communications, 4 Mbits/s Token Ring, and future

*ISDN is a set of communications standards allowing a single wire or optical fiber to carry voice, digital network services, and video. ISDN is intended to eventually replace the plain old telephone system.

technologies join voice transmissions as applications easily handled by category 5 wire.

It is important to use verified compliant components for telecommunication installations. Wire and terminals of different performance levels can look the same on the outside. Look at the Underwriters Laboratories (UL) certification to make sure that wiring is verified compliant.

More demanding applications require the use of category 5 wire. Category 5 is gaining popularity for residential use, since it provides greater resistance to interference at the higher baud rates needed by data communications equipment and will accommodate future needs for even greater bandwidth. In a four-pair category 3 or category 5 cable, each pair is twisted together to prevent induction and crosstalk interference from other pairs. The rate of twist for each of the four pairs is different to provide additional insurance against coupling. Then the entire bundle of four pairs is twisted to provide further noise rejection (see Fig. 3.5). This wire has been carefully engineered. Just 10 years ago it was nearly impossible to handle 10-MHz signals in anything but coaxial cable. Engineering advances and new wire-making technologies have made communication bandwidths up to 100 MHz possible on twisted-pair wires. But remember, good high-frequency performance comes not only from the proper wire but also from the correct connectors, jacks, and distribution blocks.

INSTALLATION GUIDELINES

Installing category 5 wire throughout a house requires more than simply selecting category 5 cabling. The cable must be run properly and all the connected components carefully chosen and installed. Improper installation of category 5 networks (not uncommon even among professional installers) is most often the cause of failures in high-speed data applications. When cables are installed, there are some important rules to follow:

- When pulling cables, avoid kinking and tugging. Constant tension should be used when pulling cable into place. EIA/TIA 568A specifies a 25-lb maximum pulling tension for category cable installations.
- Use specially designed fasteners to avoid crushing or pinching the cable.

- Avoid turns greater than 90°. Sharp angles may compromise the integrity of conductors and cables.
- For best results, cables should be installed in a home run topology, with direct connections between components.
- Avoid running cable close to external power sources. Electromagnetic interference (EMI) and radio-frequency interference (RFI) may cause data transmission problems.

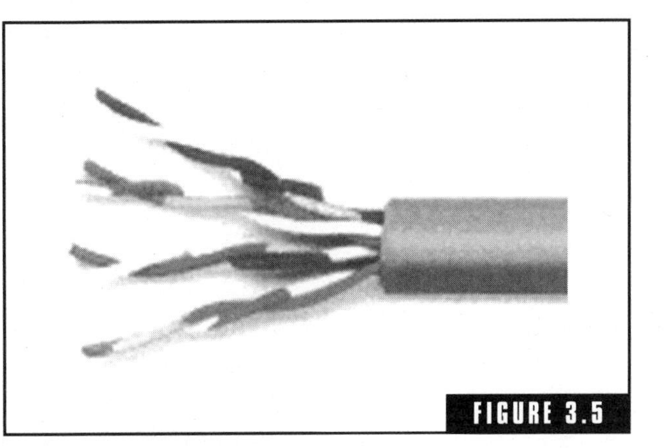

FIGURE 3.5

Four pairs of twisted wire.

- To minimize attenuation problems, keep cables away from hot or moist environments.
- Special termination procedures must be followed. Remove only as much jacket and insulation as recommended (Fig. 3.6). Do not untwist conductors more than necessary. For details, installers should consult the equipment manufacturers or EIA/TIA guidelines.

Video Cable (Coaxial)

The goal of video distribution is to route a strong clear signal, of all channels (regardless of the source), to all video destinations within the home. To achieve this goal, the distribution system must gather or create the signals; combine, condition, and amplify those signals; and finally distribute the signals to their destinations.

Video signals are really radio-frequency (RF) signals. These signals, given the opportunity, would happily fly through the air to your TVs. But this method of video distribution is frowned upon by the FCC because the FCC would rather let the licensed broadcasters handle that method of distribution. Instead, the RF signals are forced to go down shielded coaxial cables. Apart from the distribution task itself, the two most important parts of creating a video distribution system are to keep the signals inside the cables and to keep other signals out of the cables. A single coaxial

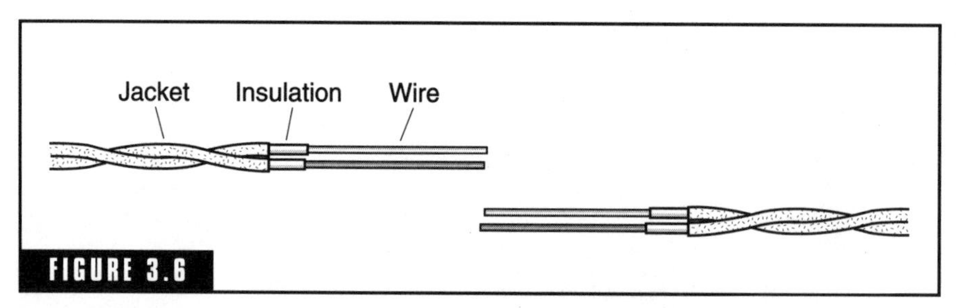

FIGURE 3.6

Remove only as much jacket and insulation as recommended for the procedure.

cable has the capacity to carry over 130 standard channel frequencies. Each channel includes video and audio components.

Contrary to RF channels, which can coexist on a coaxial cable with many other channels, *baseband* video takes the whole cable and does not include sound. Baseband video and audio is what comes out of the RCA (a.k.a. *phono*) jacks on the back of a VCR. Since it takes two coaxial cables to transport a single baseband video and audio source, baseband is converted to RF as close to the source as possible. Baseband is converted to RF via a modulator. Modulators are simple digital devices, which can be thought of as the reverse of a cable box. They have inputs for video and audio, and an output for RF. (RF is almost always an F type connector.) Modulators can be used to create "in-house" channels as well.

Planning a System

Each video distribution system includes four key pieces:

- The coaxial cable serves as a conduit for the RF signals and allows the interconnection of other key pieces.
- Splitters are used to convert the signal from a single input cable into multiple, identical signals and to distribute the same signal to two or more output cables.
- Combiners that add two or more cables create a single output containing all channels from both input cables.
- RF amplifiers "boost" the RF signals to make up for the signal loss that each of the other pieces imposes.

Coaxial cable (or "coax") includes a shield of foil and braided wire that surrounds the core. The signal is conducted through the core and the surrounding foil, and wire braid is connected to ground and used to shield the core from outside interference. This construction allows coaxial cable to conduct a wide bandwidth of signals without degradation.

RG-59 is the standard coaxial cable used for cable TV installations and the like, until recently. CEBus recommends the use of RG-6 coaxial cable for the two pairs run to each location. RG-6 uses a larger-gauge center conductor (18-gauge) and has a foil shield in addition to the braided shield, compared to RG-59. RG-6 has a lower loss at higher frequencies than RG-59. There are different grades of RG-6 cable, depending on manufacturer and cable specifications. Most consultants and sources recommended quad shielded cable, which is made of a foil shield covered by a braided shield covered by a second foil shield covered by a second braided shield. In addition, the braided shield may consist of different coverage, measured in percentage of coverage. For example, a 30 percent braid is a much looser braid made of fewer strands of copper or aluminum than a 60 percent braid. A 60 percent braid is very tightly woven. Finally, the center conductor may be made of solid copper or copper-covered steel. The copper-covered steel is used to provide greater rigidity when the center conductor is inserted and reinserted into a coaxial jack. However, solid copper is a better conductor, does not get worn off with multiple insertions, and is not as subject to manufacturing quality control difficulties during the copper-plating operation.

A *splitter* is a small device that has one input [the 75-ohm (75-Ω) load] and two or more outputs, each driving a separate 75-Ω load. Essentially they are transformers that split the power in the input signal to multiple outputs, while maintaining the 75-Ω impedance. However, every time an RF signal is split, the signal's strength is decreased.

A combiner is actually a splitter in reverse. It combines the channels on two or more separate cables onto one cable. Combiners offer a higher degree of flexibility to residential video systems. For example, a cable TV system with channels 2 through 63 can be combined with a digital satellite system (DSS) receiver so that either source (cable or DSS) can be viewed on any TV in the house. This type of "in-house" channel generation, together with the less expensive, more reliable digital modulators, is opening up many new possibilities in residential video distribution.

TABLE 3.1 Some Rule-of-Thumb Losses for Various Splitters and Cable Lengths

Device	Loss (−dBmV)
Two-way splitter/combiner	4.0
Three-way splitter/combiner	6.5
Four-way splitter/combiner	8.0
Eight-way splitter/combiner	12.0
100 ft RG-6	4.0

As the RF signal passes down the cable and through the combiners and splitters in the system, it loses strength (referred to as *attenuation*). To counter this loss, RF amplifiers are used. Ideally, the signal level at each appliance should be the same as the signal level coming in from the cable TV. This ideal is referred to as *unity gain.* Approximate losses and gains can be calculated by using the information in Table 3.1.

RF signal levels are measured in dBmV, which is a logarithmic scale of signal relative to 1 millivolt (mV). Since decibel (dB) values represent power levels and are logarithmic, they can be calculated by using simple addition and subtraction. The main thing to remember about decibel values is that if the level drops below 0 dB (into the negative decibel range), actual signal information (picture quality) is being lost and no amount of amplification will be able to recover it. In fact, amplifying a signal that is below 0 dB will usually make the picture *worse* since the noise is now being amplified and picked up. So you must ensure that your signal levels never drop dangerously near 0 dB anywhere in the distribution system. This is why the main RF amplifier is usually connected near the input side of the distribution system, so the signal is boosted early, preventing any precariously low drops.

The Head End

A coaxial cable coming into the residential system from the cable company, or an antenna feed coming in, is commonly referred to as an *in.* Outputs, or *outs,* are the cables carrying the signals "downstream" to each of the appliances or to a wall plate; these end uses of the cable are known as *drops.* Prior to designing the head end for the video distribution system, the number of ins and outs required must be known. Other inputs to the head end will come from modulators generating in-house channels. Modulators are often remotely located near the equipment where the audio/video (A/V) signal originates.

Modulators are designed to provide a range of output frequencies, and they may or may not work with a cable or antenna system. For example, some modulators provide output from 300 to 468 MHz. These modulators can be used to produce channels 37 to 64 with a cable system, but they won't work with a television that is set to receive signals from an antenna. Modulators that provide output at 470- to 806-MHz frequency are generally preferred. These can be tuned to provide channels 14 to 49 in an antenna-based system, or channels 64 to 120 in a cable-based system.

When selecting a modulator, the installer should be aware of the differences between cable and antenna systems. Cable and antenna broadcasters use different frequencies to broadcast channels 14 through 120. This is why newer televisions are promoted as *cable-ready.* In fact, what they include is a switch that allows the television to determine which frequencies to tune in order to receive any particular channel from one source or the other.

Modulated signals are then sent "upstream" to the head end over coaxial cable. So as not to attenuate the main input any more than necessary before being amplified, all inputs at the head end are usually combined separately, then combined with the main cable TV or antenna input. Unused downstream drops should be capped with a terminator, since with or without an appliance connected, the load on the system remains the same.

Although there should always be as many upstream cables coming back to the head end as there are downstream cables leaving the head end, most of the upstream cables will not be connected at any given time. Only the upstream cables that are connected to modulators should be combined into the system.

Using a preconfigured distribution panel is the easiest and cleanest head end installation. Once the number of inputs and outputs is known, select a panel with the appropriate number of ports.

Dual Coaxial Cable Wiring Systems

Probably the best option for residential installations is a cable system known as *dual coaxial communications wiring.* This system is comprised of a wall plate that includes two coaxial cable connections and an RJ-45 jack with four twisted pairs of low-voltage wiring. One coaxial cable is designated for the downstream connections to output devices, such as the televisions, VCRs, and computers. The other is designated as the upstream leg and provides for signal-generating devices such as CCTV cameras or VCR, laser disc, or computer. Several manufacturers offer dual coaxial cable systems. U.S. Tec, IES Technologies, and Greyfox Systems offer complete systems that are designed for compliance with CEBus standards. Molex and AMP offer dual coaxial cable systems under the Smart House brand name. Although Smart House is not currently designed to comply with the CEBus standard, it should work for most needs; however, installers are required to take Smart House training and certification classes to buy or install their products at this time. Keep in mind that Molex and AMP wall plates are only designed for use with other Smart House receptacles and components, while the U.S. Tec, IES, and Greyfox wall plates work with standard and Decora-style duplex receptacles.

A benefit with any of these systems is that they come bundled with telephone wiring, which connects to the same wall plates along with the coaxial cable connections. With this arrangement, users can plug a telephone jack, coaxial cable connection, or both into any wall plate. Since many locations require the types of wiring that these systems provide, the systems offer a good option for communications wiring.

The telephone outlet on these wall plates is an RJ-45 jack with eight wires. The center two pairs of conductors are used for two residential telephone lines. In compliance with the CEBus standard, U.S. Tec, IES, and Greyfox use pins 1 and 2 for an infrared repeater system. This offers a simple way to conduct infrared remote-control signals to equipment in other rooms. An infrared receiver can be plugged into a telephone jack in one room, and this pair of low-voltage wires conducts the signals to the other rooms where an infrared emitter reproduces the same signal when it is plugged into another telephone jack. The Molex and AMP systems propose that all eight pairs of low-voltage wires in their jacks are for use as four residential telephone lines, but a pair of their wires can also be used in this manner.

Fiber Optics

Many home network proponents advocate including optical-fiber cable in the wiring bundle. While there are no current applications that demand fiber media, the advantages of fiber for high-speed information distribution are notable. Fiber is immune to most sources of noise and interference and does not radiate any signal. The bandwidth of fiber cable exceeds that of any available coaxial copper cable. While fiber is currently considered rather expensive, it is actually only slightly more expensive than coaxial cable. New plastic optical-fiber cables being developed are actually less expensive than copper and are easy to handle and terminate. Most of the cost of installing fiber is due to the termination, but the cable does not actually need to be terminated until it is ready to be used. Adding (glass) optical fiber to an existing cable bundle will only add about 10 percent to the material cost and nothing to the labor cost, if it is left unterminated in the wall. This cost is not likely to be objectionable to most consumers.

Wire Routing

Plan out placements of wall boxes first. A good spot for volume controls, A/B switches, keypads, or IR receivers is near a light switch, to keep groups of switches together in a room. Keep the audio controls on studs across from the switches (one stud away) to avoid ac interference. A single-gang wall box is useful for infrequently used keypad locations, and double-gang boxes come in handy for larger keypads having more features.

Note: Some volume controls are very deep and do not fit even the deep electrical wall boxes; cutting the backs off all boxes destined for a volume control will save time and frustration later on.

Avoiding High-Voltage AC Wiring

One of the main considerations in routing low-voltage signal cable is to avoid high-voltage ac wiring. Low-voltage signal and speaker wires can perpendicularly (at 90°) cross electric wiring, but avoid running low-voltage and coaxial cable parallel or close to electric wiring. The electromagnetic field generated by the alternating current inside the power line causes noise in low-voltage, coaxial lines and speaker wire. Close proximity to electric lines can cause data errors in data lines. Common practice is to do the low-voltage wiring after the electrician has finished the prewire. Stories abound about additional labor costs and time lost because the electrician roughed-in electric wire in the holes that had already been drilled by installers running speaker wires. Make sure everyone is on the same page before lines are roughed in. Wiring on outside walls should be done prior to insulating.

There are differing views on the minimum separation between parallel runs of ac and low-voltage cabling. Some experts recommend anywhere from 6 in, and others say at least 4 ft. Most installers try to keep 3 to 4 ft of separation when possible. Frequently, wire runs to a box near a switch or floor outlet are desirable, so running along the opposite stud (16-in centers) in a bay to get to the outlet is common. If absolutely necessary, the "no parallel runs" rule can be violated for short distances, to get over a door frame or in tight locations that leave no alternatives. Never run the coaxial cable or low-voltage wiring in the same holes as the alternating current for parallel runs. Low-voltage wiring should also never go into the same wall box as ac wiring.

Special Considerations

Before any full-scale installations are begun, there are some special tools and equipment that every installer should have:

- Wire cutters or scissors
- Crimpers
- Fish tape
- String
- Cable strippers
- Connectors, based on cable types installed
- Appropriate work boxes
- Cable tester(s)
- Boxes of cable
- Zip ties
- Cable hangers or supports
- Patch panel
- Rack system (should be installed already as part of wiring closet)
- Flashlights

- Drill
- Long drill bits and/or spade bits
- Cable staples
- Punch-down tools and bits
- Stud finder
- Markers
- Hammer
- Electrical tape
- Two-way radios, if possible
- Labels

For drilling holes through wall studs, use auger bits, not spade bits. Augers have a screw tip to pull the bit through, and they cut the hole cleanly while taking out large chunks for a quick hole. A ⅝-in bit allows room for the two coaxial cables and phone or local-area network (LAN) cables for the typical TV outlet, and a 1-in bit will provide a hole large enough for all the wires in a room, leaving a little to spare. Do not use a bit larger than the 1-in. If the edge of the hole gets closer than 1 in to the edge of the stud, apply a nail plate to protect the wires from drywall screws. However, plan out your routes before you spend time getting all the holes drilled, to avoid excessive time with the drill. Holes drilled vertically in a wall stud, especially on a load-bearing wall, should be kept at least 1 ft apart. For horizontally running support studs, check local building codes.

For major cross-house runs through joists, a 2-in hole is useful. When you are drilling through floor joists, keep the hole away from the supported ends of the joists, since that is where the shear forces are located. Keep holes away from the bearing point at least 3 times the height of the joist. For an 8-in joist, don't drill closer to the bearing point than 24 in.

As a rule of thumb, never drill through laminated wood beams (lam beams). Drilling a single hole in a lam beam could prompt building inspectors to require tearing out the beam and replacing it. Again, check not only local building codes but also the attitude of local inspectors and their interpretation of the codes, before you drill lam beams (or steel beams).

A good, easy location for wire and cable runs is across the lower member of roof trusses. Connecting the ends of several wires and securing a string and a weight to the end will allow the bundle to be tossed over multiple truss members and pulled across. When you are pulling through trusses, do not pull the cables through the center of a V section—the wire will settle in the bottom of the V, and settling of the house or expansion or contraction may pinch the cables, possibly causing short circuits or cuts years down the road. When you are pulling the cable, either through holes in studs or across trusses, pull *slowly.* Pulling too fast will friction-heat the wood; the hot wood can melt or burn the polyvinyl chloride (PVC) cable jacket. And, of course, don't yank or pull the cable too tight, especially around corners or areas that might get pinched or kinked—it's not worth breaking a wire, which may not be found until it is too late.

Keeping Wires Organized

A numbering scheme for the cables is useful. A three-digit system on each cable is useful on most projects. The first digit represents the floor level. The second two numbers are sequential numbers representing the types of wires (00-20 for speakers, 21-30 for keypad, 31-50 for coaxial, etc.). Another method is to use the first digit to indicate the node location (0 for basement, 1 for media room, 2 for security center, 3 for all other point-to-point wires), and the other two digits are just sequential regardless of wire type. Keep a log of all wires run and the sequential listing grouping the cables by room.

The Automation Closet

The best run, most organized wires leading from wall plates in rooms throughout the house all have to lead to one central location—the automation closet. This space should be planned for in the blueprints of new construction, can be added to existing structures, or can be improvised from unused or sacrificed space in an existing home.

Key considerations to locating an automation closet are both aesthetic and practical. As shown in Fig. 3.7, the automation closet will have items such as

- Distribution panels that route audio and video signals throughout the home

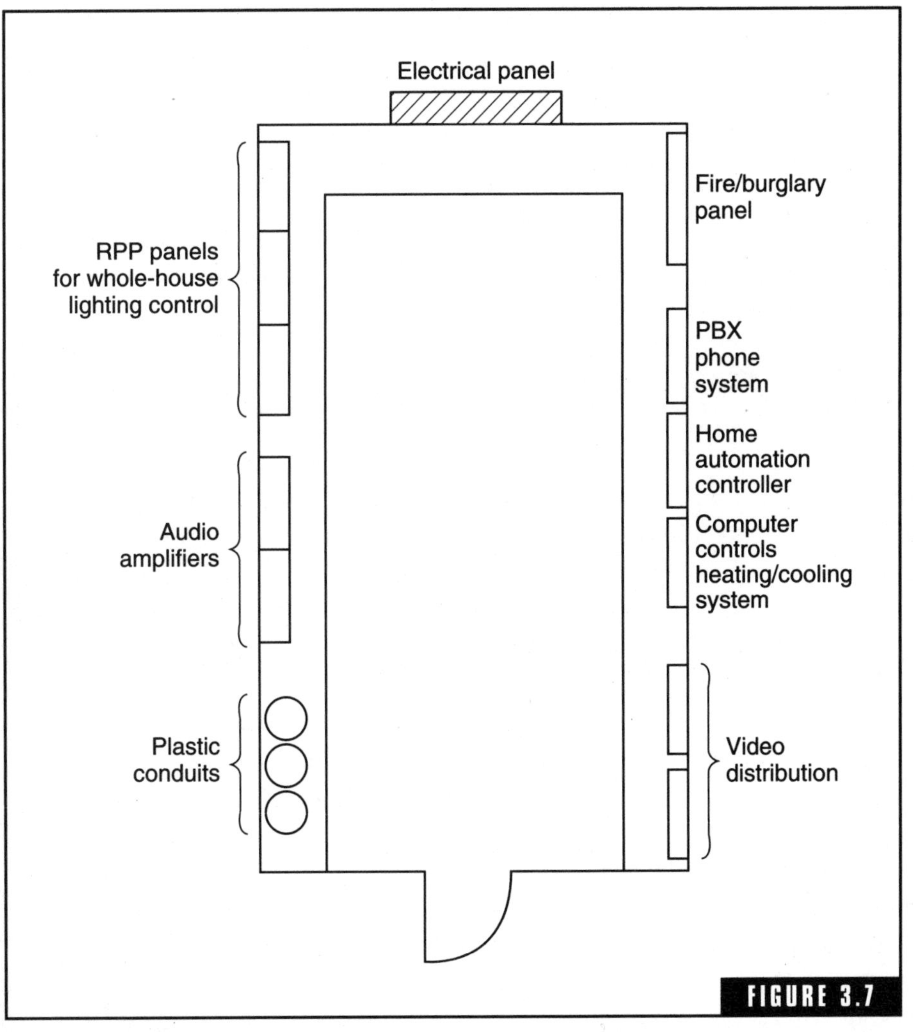

FIGURE 3.7

Example of a typical automation closet.

- Enclosures that house dimmer modules for the whole-house lighting control system
- Central processing units (CPUs) for various subsystems
  1. Home security
  2. Automation
  3. Energy management
  4. Telephone

The homeowners will want this room situated in a low-traffic area where it will not be conspicuous. Also because of the electronics, the closet must be in a controlled environment. Attics are subject to extreme temperatures that would damage the sensitive gear. Garages or basements with workshops are no good due to high levels of dust, while unfinished basements are usually too damp.

Preferable locations would be closet space inside the living area of the home, finished basements, or heated and air-conditioned garage space. Normally, a 5-ft by 12-ft space will be adequate for the automation closet. A room built for this purpose in a finished basement is ideal because the room is easily accessible, out of the way, and close to key equipment such as the HVAC system.

Remember: Allow space within the automation closet for future expansion. As technology continues to develop, new systems will most likely be added to the home.

Attics, Ceilings, Outside Walls, and More

Although it is not recommended that electronics designed for indoor use be mounted on outside walls (such as audio speakers), sometimes there is no other choice. Situations like this call for some ingenuity and planning. For example, mounting speakers on the slope of a cathedral ceiling of considerable depth may require the displacement of some insulation to accommodate the backs of the speakers. The main concern is the vapor barrier, since the installation has to ensure that the backs of the speakers never get damp. To make room for each speaker, cut the vapor barrier on the room side of the insulation. Then tape a larger piece of vapor barrier behind the cut hole to ensure that any moisture between the insulation and the plastic will run on the other side of the speaker. With this installation there is still insulation on the outside with the patched plastic barrier, then the speaker.

Ceiling-mounted speakers in which the attic is above the ceiling are another tricky installation. Custom-fabricated insulated "boxes" built between ceiling joists in the attic space above the speaker locations will help. In new construction these boxes should be built after drywall installation and before the insulation is put in. Using 2-in foam insulation panels, or blueboard, cut end panels to the height of the ceiling joists and run them between the joists. Run the speaker cables into this newly created chamber, and nail a top cover of the same material over the joists and end panels. Seal all joints with caulk to make sure insulation particles and dust will not get into the speaker chamber. After completion of the house, cut in holes for the speakers just as for a wall-mounted location. This solution will provide a sealed chamber for each speaker without heat loss.

Empty Conduits

The most compelling reason to install wiring inside the walls of new homes is to provide access for future needs. In many cases, installing empty conduit between likely control points (such as wall switch or keypad locations), subsystem locations (such as media equipment, security, and HVAC equipment rooms), and wiring distribution points will provide access for any future wiring requirements. Conduit is more expensive to install, however, and the cabling will still need to be pulled later. Providing conduit raceways between floors and from the attic to basement or crawlspace is still highly recommended.

Raceways

Frequently, even in prewired homes, there are wire runs that cannot be hidden inside walls, ceilings, or floors. In this instance products known as *raceways* are used to manage those runs. In addition to runs of house wiring, speaker wiring, and low-voltage and command wires, raceways offer an easy, fast, and flexible way to manage the tangle of wires which normally come with a home office.

Extruded nonmetallic raceways in decorative baseboard or chair rail design can do much to eliminate the tangle of wires and cables associated with computers and various communications devices beneath desks, behind book shelves, and even under carpets. Raceways are usually equipped with access panels—covers that snap on and off when it is necessary to add new wiring or to reposition outlets.

Wire management systems such as raceways have been popular in commercial office space for years, but today the practical advantages of raceway technology are being combined with designer looks to complement any home office. These raceways can be installed in new construction and are ideal for renovations.

Device plates, which can be easily installed anywhere along the raceway, accommodate power receptacles, phone jacks, and outlets for data, coaxial cable, security, and home automation cabling. This allows homeowners to place equipment where they want it, instead of organizing and basing decorating plans on the position of in-wall outlets.

Installation is very simple. Just mount the base on almost any surface (plaster, concrete, brick, drywall, or paneling will easily accommodate raceways) and run the wires. Covers and fittings snap on over the wires and the base.

Wiring Ducts

Another product available for routing and managing wire runs is wiring ducts made of high-impact and rigid PVC. Solid, standard slotted, and high-density slotted channel wiring ducts are the three most common styles available. More than 26 different sizes of ducting are available in standard 6-ft lengths. Standard colors are normally gray and white; however, custom colors are possible based on minimum volume requirements. Variations of wall ducts are as follows:

- Solid wall duct, designed for straight wire runs where breakouts are not required.
- Slotted wall duct with break-away fingers which provide additional access for wire leads. Each section of the duct is provided with two score lines. The upper score line is used to break away fingers when a greater opening is required. The lower score line allows the sidewall to be smoothly cut for joints and tees.
- High-density slotted wall duct designed with narrow fingers, giving a 2:1 finger ratio compared to standard slotted wiring duct. The narrow fingers reduce fanning of the wires to the terminal blocks for a neater appearance and to accommodate more compact designs and components in control panels, communications closets, and other applications.

Fiber-Optic Enclosure Systems

Fiber-optic enclosure systems consisting of connectors and protective channels are designed to route and protect fiber-optic cable. Enclosure systems prevent accidental cuts, kinks, and associated signal loss.

Connectors should provide a 2-in bend radius to ensure effective signal transmission while utilizing minimum space.

Fishing Wires

When it is necessary to fish wires between floors after construction, there are some tips to make the installation easier. When the ceiling material is smooth, such as with sheetrock, follow these steps.

1. On the lower floor, use a stud sensor to locate the two joists between which the device will be placed.
2. Cut a hole a least 2 in in diameter between the located joists; then measure the distance from that hole to the outside wall parallel to the joists.
3. Upstairs, measure the same distance from the same outside wall, in order to be between the same joists, against an inside wall that is perpendicular to the outside wall.
4. Pull up the carpet and drill a hole between the tack strip and the molding. Tie off or use a stop ring on the fishing chain, and drop it in the hole. It will pile up on top of the ceiling.
5. Back downstairs again, open the ceiling and locate the pile of chain.
6. Once you have the chain, attach to it a minimum of 10 ft of pull cord and the wire or cable to the cord.
7. Upstairs, the chain, the pull cord, and the wire (or cable) can be pulled into position.

Many customers are already aware, if not convinced, of the need for better wiring in their homes. The appeal of prewired homes is proving to be far greater than most skeptics had imagined. The incremental cost is very low compared to other aspects of construction and custom home electronics, and the inside wiring is one of the most permanent parts of a home.

Notes

Notes

Notes

CHAPTER

4

Automating the HVAC Systems

The heating, ventilation, and air conditioning (HVAC) system of a home is the single most costly system to operate, accounting for approximately 70 percent of the home's total utility bill. It is also the system most responsible for the comfort of the homeowners. Zoned heating and cooling systems allow the temperature of each room to be controlled individually. Zoning helps eliminate hot or cold spots (rooms) by utilizing a series of motorized dampers and thermostats that work independently. Automated zoning increases the HVAC efficiency while also cutting energy use. This translates to big money savings on utility bills.

Not too many years ago, for people upgrading their HVAC system or buying a new home, energy savings were key considerations when it came to air conditioning and heating. Still a prime concern, energy efficiency is now joined by the comfort factor as the two major aspects influencing the purchase decision. Once the benefits of an automated and zoned HVAC system are fully explained to the buyer, it becomes an easy sale. The benefits realized by the homeowners and contractors make automating the HVAC system a practical necessity.

Zoned Control

It is virtually impossible to keep an entire house at a consistent temperature without zoning. Since warm air rises while cold air falls, rooms above the ground floor average 6°F to 10°F warmer while rooms below the first floor average 6°F to 10°F cooler than those on the ground level. A system equipped with a single thermostat keeps the temperature balanced in the room where the thermostat is located. The downside to this type of system is that the single thermostat is not aware of temperature changes in other rooms of the house. Zoning maintains a consistent temperature throughout the house by providing multiple thermostats and thus different levels of air distribution to different areas of the home.

Dividing a home into multiple heating and cooling zones, as seen in Fig. 4.1, separates that home into areas which share common heating and cooling needs during specific parts of the day. Instead of sending the same amount of heated or cooled air into all the rooms each time the furnace or air conditioner is turned on, the system sends conditioned air only to the zones that need it.

Occupied areas can be kept at an ideal comfort level while unoccupied areas are kept at more energy-saving temperatures until shortly

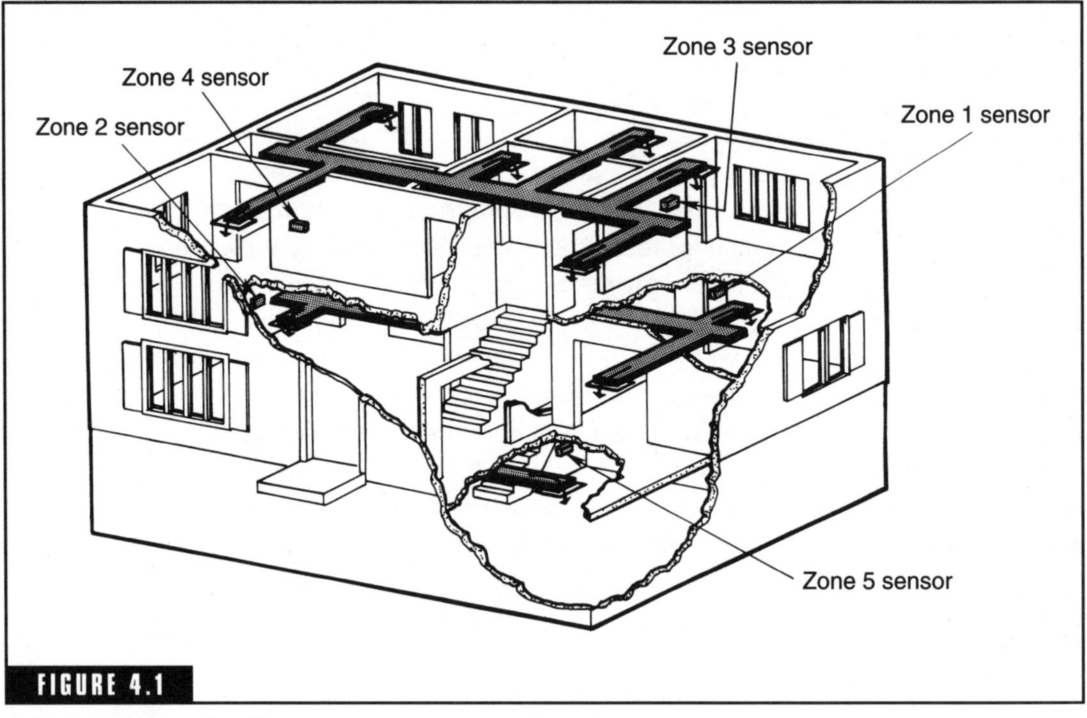

FIGURE 4.1

Multiple heating and cooling zones.

before they will be occupied. For example, unoccupied bedrooms, baths, and studies on the second floor can be kept warmer (or cooler) than the occupied living room, kitchen, and dining room. An hour or so prior to bedtime, the upstairs will be adjusted to more comfortable temperatures, and the downstairs will go into a predetermined energy-saving mode.

Most homes can be zoned according to room occupancy, but unique exposure factors may require a different zoning strategy. A room with large amounts of glass facing south or west will have more heat gain than other rooms in the home. A separate zone might be required for that room alone.

Zone-controlled systems are especially effective in homes that have multiple levels, sprawling designs, large glass windows, or large open areas such as an atrium or solarium. Homes with finished basements, attic spaces, and additions are also good candidates for zoning. Almost all forced-air systems can be converted for zone control. A control panel, thermostats, dampers, and the proper ductwork are all it takes. Zone-controlled systems offer a cost-efficient alternative to dual air systems that require separate air conditioners and furnaces for different areas of the house.

Thermostats and Humidistats

As more and more homes are equipped with zoned systems, hydronic heating zones, and other systems that require multiple thermostats, the potential control problems and solutions grow. Communicating thermostats, as seen in Fig. 4.2, meet the challenge and provide new opportunities in comfort and savings. With each zone having its own thermostat, home automation provides the means to regulate each zone separately or collectively or in any combination.

Several different approaches are available for controlling thermostats. One of the control choices involves thermostats that listen and respond directly to X-10 signals. Different unit codes sent to the thermostat(s) result in different preprogrammed set points. For instance, unit 2 ON may result in a set point of 72°F, while unit 2 OFF changes the set point to 68°F.

Another possibility involves the use of contact closure outputs. Originating from any source, including an X-10 universal module,

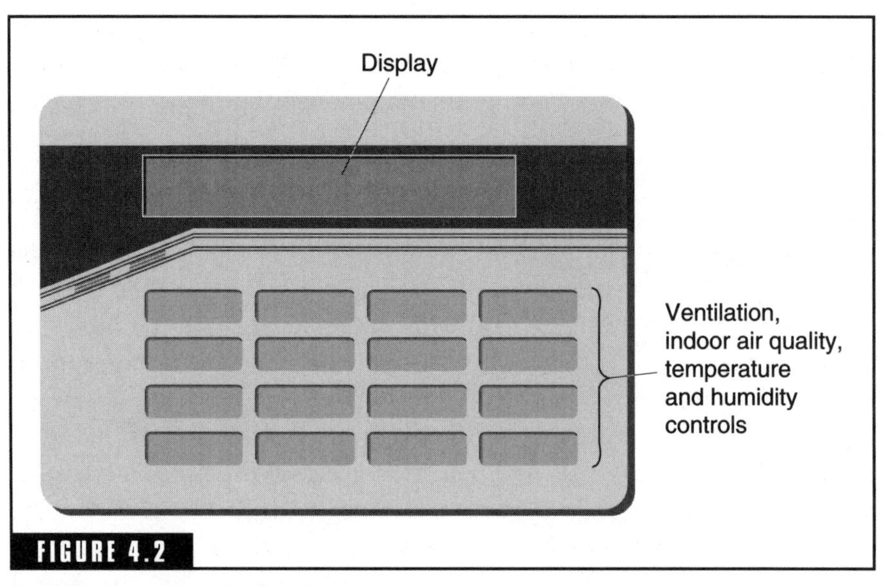

FIGURE 4.2

Example of a communicating thermostat.

contact closure outputs are used to trigger preprogrammed set points. As an example, a relay output from a home automation system, when open, sets the thermostat to 72°F, and sets the temperature back to 68°F when closed.

The most sophisticated choice is a true communications link between the HVAC system and a home computer. The computer is directly linked to the thermostat, resulting in an image of the thermostat on the computer screen. Homeowners, installers, and technicians can make and observe any changes from the computer just as if they were working directly on the thermostat. Almost all major suppliers of automation systems include this type of thermostat as a part of the system.

Considering the Needs of the Homeowners

The thermostat industry established most conventional programmable thermostats to offer four different set points per day. This standard was based on what industry studies had shown to be the lifestyle needs of an average user. For example, typical set points for these conventional programmable thermostats would start with a setting of 72°F in the morning while people are getting ready for work. The system would then be reset to 62°F during the day while the house is unoccupied. When the user returns home, temperatures are readjusted to 70°F for the evening. At bedtime, an economical 68°F is established while the user sleeps. However, today's reality includes home offices, erratic schedules, full-time parents, and other occupancy schedules that can make the standard programmable thermostat with its rigid scheduling useless.

Communicating thermostats can be adjusted in person with up and down arrows, just as any electronic thermostat. These thermostats add the dimension of letting homeowners call ahead from their offices, phone booths, or mobile phones to change the temperature at home. Being tied into the home automation system, a change in thermostat setting can communicate to the lighting, security, or entertainment subsystems, or almost any other device in the home. Based on its programming, the thermostat can use other control events in the system to decide what temperature level to maintain. The end result can be no more wasted dollars heating or cooling an empty house and no more coming home to an uncomfortable house.

Protocol

CONTACT CLOSURE

X-10 users can use either single- or multiple-contact closure thermostats or a direct X-10 thermostat. With the models, the thermostat can connect to X-10 universal module(s) to select between preprogrammed set points. Using binary logic, one contact closure gets two set points, and two contact closures get four set points. An X-10 system is not required, since any source of relay outputs or contact closures can be used, meaning control can be achieved from hardwired relay outputs, security systems, motion detectors, and other non-X-10 protocol devices. The main disadvantage concerns remote operation. Not just any temperature can be selected; only preprogrammed set points are offered as choices.

X-10 DIRECT

An X-10 direct thermostat actually listens for X-10 signals, just as a lamp or appliance module does. By remotely selecting a unit code and either an on or off command, temperatures can be adjusted in 2°F increments; X-10 direct thermostats offer great functionality for the die-hard X-10 user, since so many remote set points are possible and will retrofit with existing thermostat wires in many applications.

However, four separate parts make for a cumbersome installation and could, without proper installation, be inadvertently turned off by your neighbor, line noise, or some other X-10 glitch.

TRUE COMMUNICATIONS

True communicating thermostats are those linked with a computer or software-driven system of some type. This is the thermostat communication strategy of choice for upper-end automation systems. The user can see an image of the thermostat on the screen or touch pad and can adjust and manipulate the thermostat just as if he or she were standing in front of it. The system software allows many control features above and beyond any stand-alone thermostat, conventional, contact closure, or X-10 direct. Simplified scheduling, event-based control, historical and energy use reports, alarm modes to alert for system malfunctions, and many other features become possible.

Several types of thermostats are now on the market that will work with virtually any type or brand of HVAC equipment and home automation system. Whether serial port, X-10, or contact-closure-driven, these automated thermostats carry out their tasks with an added bonus: If the automation system shuts down, these new communicating thermostats revert to a stand-alone mode and continue to operate the HVAC without missing a beat.

A communicating thermostat has some obvious benefits, two of which are its easy-to-use scheduling capabilities and its call-ahead feature, allowing remote control of the system so the homeowner can adjust the temperature before arrival. Conventional programmable thermostats are sometimes difficult to program. Automated thermostats can be programmed through the home automation system by using fill-in-the-blank-software on a 15-in color monitor. Programming and scheduling become fun. And fun, easy-to-use scheduling usually translates to thermostats that get scheduled, not forgotten.

Interactive thermostats within the home automation environment can be combined with the security system. When the security system is set to the away mode, all the thermostats automatically adjust to energy-saving temperatures. High-efficiency air filtration can benefit from automation, which can operate as a runtime accumulator and alert the user when cleaning or servicing is needed.

Energy recovery ventilators (*ERVs*) pull polluted inside air out of a home, while bringing fresh air in. Prior to being exhausted, the outgoing air is run through a heat exchanger. This process recovers most of the energy contained in the outgoing air. An ERV under the control of an automation system knows the outside temperature and humidity, as well as conditions inside, and can run at full tilt when it makes sense, and not at all when outside air is undesirable.

Temperature Control

The sophisticated controls on high-technology thermostats offer heating or cooling within 1°F of the set point. Seven-day programming provides the flexibility to program different schedules every day of the week and thus save energy automatically. A Hold key holds the current temperature setting for a selectable period (up to 7 days). Automatic changeover improves comfort, especially in climates needing heat by night and cooling by day. Optional temperature sensors allow different temperatures to be averaged out in different areas of the home, providing a more comfortable and consistent temperature throughout.

Sensors

An integral part of home automation control of HVAC systems centers on temperature sensing. Outside temperatures, inside air temperatures, appliance temperatures, or any other temperature-related item has to be converted from real-world analog measurements to a digital form readable to the software. Two products that are readily available to convert temperatures from real-world analog measurements to digital form are commonly utilized. While appearing physically similar, these two semiconductors are actually quite different. One device, the LM334, is a current source that possesses a linear dependence on temperature in its operating characteristics. This dependence is calibrated (and is directly related) to temperature, making it simple to perform a conversion of the digital information to usable temperature information. The second device, the LM335, is designed as a temperature sensor with a similar dependence on temperature. Either device is capable of yielding temperature measurements in either the Fahrenheit or Celsius measurement system by means of a conversion formula. This means that the selection of which device to use is more an issue of availability and accuracy than one of which measurement units are desired. Each device is available in myriad operating temperature extremes and accuracy tolerances.

CONNECTION AND BIASING

Connecting these two devices involves dealing with subtle differences since the devices are of a fundamentally different design. Since the LM334 is a current source, it must be programmed and a load impedance supplied to develop an output voltage which can be measured. Figure 4.3 demonstrates how this step is accomplished. Note that the third lead of the LM334 is not used. It can be used to fine-tune the output voltage with respect to temperature, but a similar effect can be achieved with a simple offset in software.

The LM335, on the other hand, functions as a zener diode and must be reverse-biased in order to generate a temperature-dependent output voltage. No load impedance is necessary, but the LM335 does require a current-limiting source impedance to place it in the proper operating mode (see Fig. 4.4).

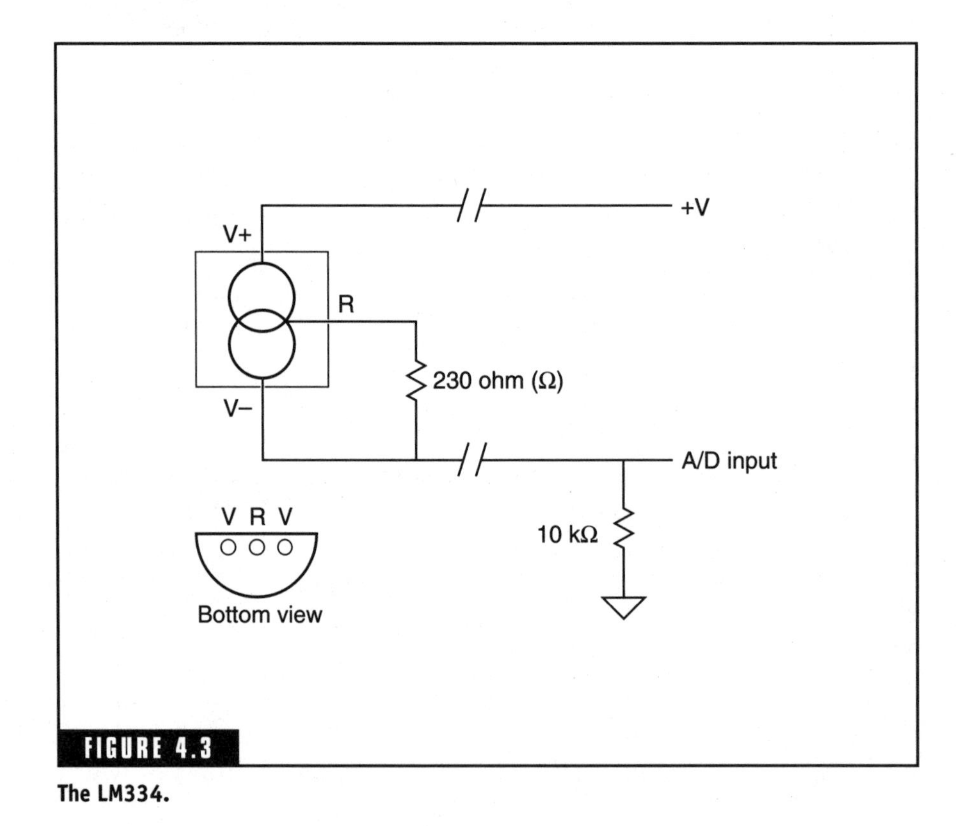

FIGURE 4.3

The LM334.

In highly noisy environments, it may be necessary to filter the analog output of the device. Adding a 0.01-microfarad (0.01-μF) capacitor from the A/D input to ground is a logical solution. In both cases, the resulting output voltage can be directly fed into the analog-to-digital converter for measurement. Also, the power supply voltage used to supply the devices must be equal to or less than the maximum allowable input voltage for the analog-to-digital measurement system to avoid damage. Usually this supply voltage is 5 volts (V), as assumed in Figs. 4.2 and 4.3. For other voltages, the biasing resistors must be adjusted to ensure that the devices are functioning in the proper operating range. It is useful to note here that the operating currents used for the devices should be kept at the lower end of the allowable limits, to reduce any possibility of self-heating. Self-heating will degrade the accuracy of any measurements.

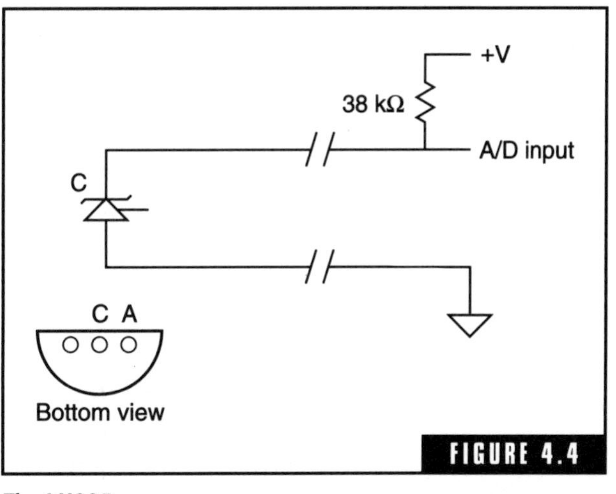

FIGURE 4.4

The LM335.

Converting to Digital Form

The analog-to-digital conversion that is performed on the input voltage will yield a digital number, which depends on two variables. First, the size (number of bits) of the converter dictates how large the resulting digital number will be. Typical converters can be 8 bits, 10 bits, or 12 bits, and will yield maximum *counts* of 255, 1023, or 4095, respectively. This means that for an input voltage of 0 V, any of the converters will yield a count of 0. But for an input voltage equal to the maximum allowable for the converter, the values of 255, 1023, or 4095 will be returned. Any voltage in between these limits will be represented by the converter as a fraction of the maximum number in the same proportion as the input voltage is to its maximum. So for an input voltage of 20 percent of maximum, the count returned will be 20 percent of the maximum count.

The relationship between temperature and input voltage needs to be understood in order for the digital information to be truly meaningful for temperatures. When configured and biased properly, both devices have an output voltage directly proportional to absolute temperature. The relationship is equal to 10 millivolts per kelvin (mV/K). In practical terms, at a temperature of absolute 0, the output voltage will be 0; at a temperature of 1 K (absolute 0 plus 1 degree) the output voltage equals 10 mV (0.010 V). Since this relationship is a linear one, dividing the voltage by 0.010 will yield the output temperature. Since temperature in kelvins is not the most common reference used, most people will need to convert kelvins to either degrees Celsius or degrees Fahrenheit to make the information more useful.

To convert kelvins to degrees Celsius, simply subtract 273 from the above calculation. The freezing point of water is 273 K, which is equal to 0°C. If the calculated input voltage equals 2.73 V, then the temperature of the sensor is 0°C. For output in degrees Celsius, that is all there is to it. To convert readings to degrees Fahrenheit, multiply the calculated Celsius output by 9/5, then add 32; the result will equal the output in degrees Fahrenheit.

OUTDOOR TEMPERATURE SENSING AND FROST CONTROL

Optional outdoor sensors provide a convenient method for checking the outdoor temperature. They are more convenient than a thermometer because the outside temperature displays on the control center with electronic accuracy. Outdoor sensors can also be used to apply *humidity-reset* technology to the humidifier. As the outdoor temperature drops, the controller automatically lowers the humidity setting to reduce or eliminate frost or ice buildups on windows during the cold days of winter. It optimizes the humidity level for comfort and health, while minimizing the damage that excess moisture can cause. The frost setting can be fine-tuned based on home construction. One setting is used for homes with tight construction where there is too much condensation on the windows in cold weather; another setting is used for homes with older, less energy efficient construction where there is little condensation on windows.

INDOOR AIR QUALITY CONTROL

Electronic media air cleaners quietly and efficiently remove allergy and asthma-causing indoor air contaminants. The control center automatically indicates when air filters need cleaning or replacing. Normally after 720 hours (30 days, 24 hours/day) of fan runtime, the control interface unit will display a message indicating the filter should be changed. The 720-hour period starts over when the displayed message is canceled. The fan feature of the system circulates air according to a schedule, improving comfort by reducing air stagnation.

VENTILATION CONTROL

The system controls a heat recovery ventilator or fresh air damper, continuously providing fresh air to the home and removing stale air. The ventilator function is activated in response to any of the following conditions:

- A ventilation request by the homeowner
- A high carbon dioxide (CO_2) condition (when a CO_2 monitor is configured and installed in the home)
- High humidity while in the heating mode

USER INTERFACE

The user interface is usually a slim panel that is mounted on or in the wall of living space. Messages are relayed onto a display very often back-lighted for easy reading and programming. Either manual or automatic heating-cooling changeover is accomplished with a simple switch.

CONTROL MODULE

The control module is the brain of the system. Remotely located near the heating and cooling equipment, most units can be mounted on any flat surface or on the exterior of the heating appliance itself. The user interface and control module pass messages via communications technology accomplished through the same four wires needed to install a standard thermostat. All wiring is 24 V.

COMPATIBILITY

Many of these systems are compatible with more than 99 percent of residential heating and cooling equipment on the market and can be installed in virtually any home. The remote module located near or on the heating and cooling equipment allows for easy installation with the wires all in one location.

Most control systems of this nature are set up to handle the elements previously discussed. However, if the home is not equipped with whole-house ventilation, high-efficiency electronic air cleaning, humidifiers, or dehumidifiers, these control systems can initially be used for temperature control only. The home will be ready for system expansion later, without installing any new controls. If the home already has air conditioning, there is a high probability that no extra wires will have to be run to set up the basic system.

Remote sensors can usually be used in three applications:

- *Remote sensor only:* In this application, a sensor is mounted in the living space. The control center adjusts temperature based on the sensor location only.
- *More than one remote sensor:* In this application, two or more remote sensors are located in a given living space. The control center manages temperature based on an average of the locations.
- *Remote sensor network:* In this application, several remote sensors are mounted throughout the living space and arranged in series or parallel, forming an averaging network. The control center manages temperature changes based on the inputs from the averaging network.

Electronic Duct Dampers

Electronic duct dampers (Fig. 4.5) serve as the gateways to allow or stop airflow from the HVAC system to various rooms throughout the house. Apply 24-V alternating current to these dampers, and the normally open dampers close, shutting off all airflow. Apply the same current to a normally closed damper, and it will open, allowing airflow to resume. If power is removed, a spring will automatically return dampers to their original position. Most in-line dampers are equipped with a specially designed seal that ensures no air leakage when closed, to prevent energy loss and wind noise.

When used with multiple-zone thermostats or control panel, a two-conductor wire (18- to 22-gauge) is run from the zone thermostat or control panel to the damper. The thermostats or control panel automatically

FIGURE 4.5

Typical electronic duct damper.

opens or closes the dampers to maintain the desired temperature within each zone. When the home automation system includes carbon monoxide, smoke, and/or fire detection equipment, the dampers can be programmed to close when one of these devices sounds an alarm. In the case of a fire, e.g., the dampers would close immediately, preventing deadly smoke from being circulated throughout the home.

Installation simply requires cutting out an appropriately sized section of the duct and inserting the damper in its place. Use duct tape or strap it in place, and run the wires back to the zone thermostat or control panel. Power consumption to maintain the damper in position is very low.

Duct Boosters

In many larger homes, long duct runs decrease the airflow into certain areas of the house. Even in homes with properly designed heating and cooling plants and correctly sized ducts, there are likely to be some rooms that don't cool off or warm up enough.

Turning up the temperature for the whole house just to compensate for rooms that are too cold, or lowering the thermostat to cool those rooms that are too hot, overheats the rest of the home, annoying the residents. This behavior also wastes energy and costs the homeowner money. Raising the thermostat by just 4°F will increase the average utility bill by $35 per month. Lowering the cooling temperature by the same 4°F will increase the average utility bill by approximately $50 per month.

In-line duct boosters, as seen in Fig. 4.6, can be used to have the airflow to individual rooms or areas of the home boosted by up to 80 percent with no noticeable additional noise. Increase ventilation to selected areas when there are guests, or equalize temperatures in rooms that are too hot or too cold.

Installation is easy. Simply cut the flexible or rigid ductwork where the booster is to be installed. Insert the booster in the duct, and seal the duct with duct tape and nylon strapping.

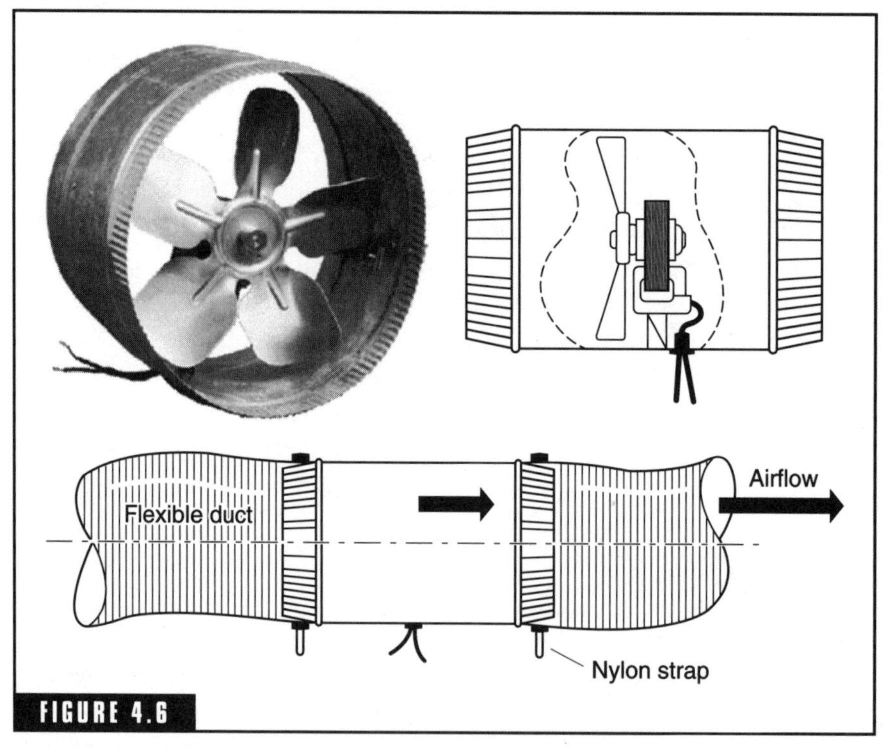

FIGURE 4.6

Typical in-line duct booster.

Duct booster fans can be wired directly to a wall switch for manual control or to an HVAC system (a 110-V relay may be required). These boosters can be used within an X-10 control environment in conjunction with a 2001 X-10 appliance module (plug-in) or with a 2250 Leviton fixture relay module (in-line wiring) to obtain remote X-10 control of the booster.

Air Circulation

Air circulating into, throughout, and out of a home performs a multitude of tasks. Replacing stale air, evaporating moisture, and cooling are all accomplished by moving air. The various pieces of equipment used to circulate air for different purposes are all capable of being automated to some extent.

Attic Ventilator

Power attic ventilators, regardless of placement within the attic, can be automated with the addition of thermostats and humidistats. Although independent of whole-house automation, these devices are indeed automated, since their operation requires no thought or action on the homeowner's part. When humidity levels or temperatures rise above a preset limit, the fan will be triggered to operate until temperatures and/or humidity levels are again acceptable.

Whole-House Fans

A whole-house fan is a simple and inexpensive method of cooling a house and exchanging stale air. The fan draws cooler outdoor air inside through open windows and exhausts hot room air through the attic to the outside. The result is excellent ventilation, lower indoor temperatures, and improved evaporative cooling.

A whole-house fan can be used as the sole means of cooling or to reduce the need for air conditioning. Outside air temperature and humidity dictate times when the whole-house fan would be more favorable than air conditioning. If both methods of cooling are present, a seasonal use of the whole-house fan (during spring and fall) may yield the optimum combination of comfort and cost. Primary residences that are closed up during family summer vacations as well as vacation homes, which are unoccupied for lengthy periods during summer months, can register temperatures well above 100°F. Air conditioning systems take considerable time to lower those high temperatures to comfortable levels. A whole-house fan can quickly exhaust the 100°F+ air to the outside while replacing it with air 10, 15, maybe 20°F cooler, depending on outside temperatures at the time. Once this is accomplished, homeowners can switch over to the air conditioning system to finish the job.

Controls may be simple on/off pull or wall switches, multispeed rotary wall switches, or a timer, which automatically shuts off the fan at preselected time intervals. Whole-house fans can be wired into home automation controls to function at various predetermined settings or can be triggered manually. However, as with attic fans, whole-house fans can be equipped with thermostats and humidistats to automate their functions in a stand-alone format.

There are two basic whole-house fan designs to choose from: direct-drive and belt-drive. With either one, you can get a variety of speed controls, including single, triple, and variable. Direct-drive fans are easier to install than belt-drive because they don't require attic floor joists to be cut. Although belt-driven whole-house fans may be a little more work to install, because joists need to be cut during installation, they offer increased circulation and quieter operation.

Many whole-house fans include dampers and/or louvers, much like the example in Fig. 4.7. Typically these louvers operate automatically whenever the fan operates. However, when connected to an automation system, these vents are controlled via signals from other communicating devices such as smoke detectors.

Windows, Skylights, and Doors

Many forms of intelligent controls have been utilized in commercial and industrial applications to open and close windows, doors, and gates for many years. This technology is now available in home automation systems and is practical for many consumer uses. Some of the most common controllers include mechanical window openers, motorized window coverings, access entry devices for exterior doors, and both access entry and motorization for gates.

MOTORIZED WINDOWS

The benefits of remote-controlled window openers are obvious, but aside from convenience, there are other benefits to this technology.

FIGURE 4.7

Typical dampers and louvers.

The system can open the windows as an alternative to using the HVAC system and save money in terms of heating or cooling needs. If the home automation network includes moisture sensors, the system can automatically close windows when it begins to rain, and electric locks allow the windows to be automatically locked when the homeowner arms the security system.

Motors for these windows are commonly small and, as seen in Fig. 4.8, attached to the mechanical opening mechanism. These motors are connected to standard electric power and use an interface device to control them. Similar motorized control equipment is also available for skylights. These motorized windows can be controlled as well by handheld remotes, wall-mounted keypads, telephones, computers, and televisions. When an automation controller (such as a PC) is in use, these windows can be automated to open or close at preset times, or when outside air is desired, or as part of a "scene." Windows can also be operated automatically in conjunction with other devices, such as when motion or smoke is detected.

ACCESS ENTRY DEVICES

Access entry devices replace peepholes as a means of letting homeowners see who is at the door or gate from any nearby television. The door or gate can be unlocked by using preset codes from a keypad or any convenient telephone inside or outside the home.

MOTORIZED GATES

Much like automatic garage doors, motorized gates provide homeowners with convenient remote control. Gates can be opened and closed

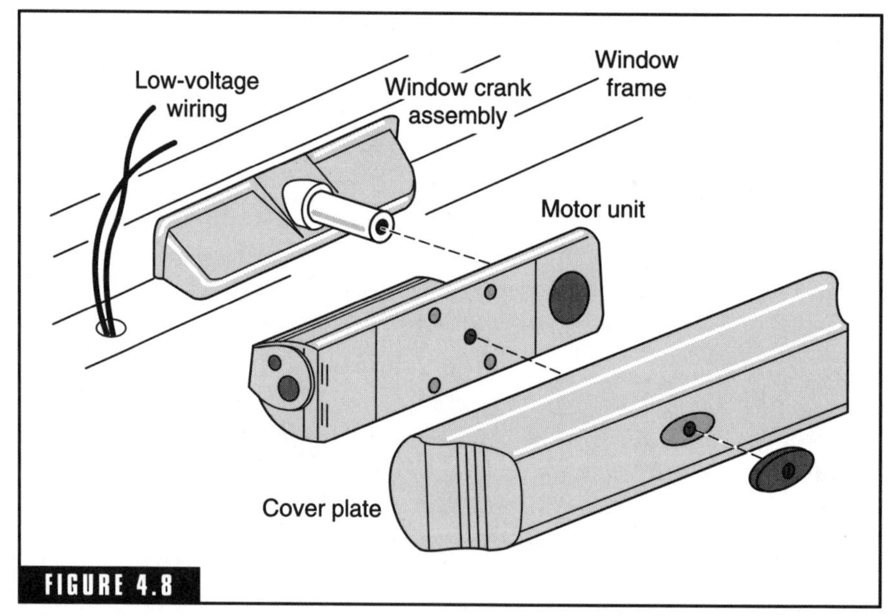

FIGURE 4.8

Example of a motorized window.

from inside the home or vehicle with small battery-operated remotes. Sensors can be added to further automate the system. Sensors are attached to the vehicle so that the gate or garage doors can automatically open or close when the vehicle approaches or departs.

MOTORIZED WINDOW COVERINGS

Motorized window coverings and drapes provide convenience through remote controls. They can also be set up to automatically open or close at certain times, such as opening when the homeowners wake up in the morning and closing when the sun goes down or when a home theater is turned on.

A standard electric outlet installed at the base of the window usually provides the power for a motorized window and its interface. A dual-gang box is often used to provide a location for both a receptacle and an interface module. Some controls for window coverings are designed to be hardwired to standard household power or a low-voltage supply source. Installers need to be aware of these variations and verify the method required for the product that they intend to use before running wires.

Although drapery controls are the simplest, motorized controls are available for just about any type of window covering including vertical and venetian blinds and shutters as well as draperies. A variety of electric motors are available to open and close drapes. Some of these attach directly to the rail on which the draperies ride, and others attach to the cord that is used to open and close the drapes manually.

Vertical and venetian blinds offer two options. A single motor can be used to open and close them like draperies. Many blind controls also include a second motor that can change the angle of the blinds to alter the amount of light being admitted or shut it out altogether. Shutters also offer two options. Depending on the design, some shutters are decorative as well as functional and are constantly exposed. An electric motor puts the shutters into either an opened or closed position. Another type of shutter rides on tracks, retracting vertically or horizontally out of sight.

Most motorized window coverings run on standard 120-V ac electric power. Power supplies are utilized in some applications to provide low voltage for motors and controls. Some systems include a factory-installed interface to facilitate addition to the home network, while others require an installer to add the necessary interface(s) to enable control through other resources.

PATIO DOORS

Virtually any sliding patio door can be converted to a motorized remote-control entrance. The benefits of this type of door are numerous. It provides access for the physically challenged, makes entering or exiting the home easier for elderly persons, and provides access to children, who do not have to carry keys. Security is increased by preventing the door from being accidentally left open or unlocked. This application can also be used to prevent young children from entering swimming pool or spa areas through open patio doors.

Self-charging battery backup systems enable the unit to continue working even during a power outage. Remote controls can be handheld or wall-mounted. Optional combination keypads are also available for external access.

Main Entry Doors

Although used for years on commercial and industrial buildings, automatic main entrance doors have not yet become popular with the general population. However, a segment of our population, which deals with special access needs, has benefited from this technology. People confined to wheelchairs have always had a difficult time entering and exiting their own homes. Extra-wide doorways equipped with hydraulic door openers are meeting the needs of these people today.

Easy to install, these hydraulic motors are usually placed at the top of the door frame with the arm attached to the door. As the arm is extended or retracted, a rubber roller (in direct contact with the door) pushes the door open. After a preset time, the door will automatically close. Control is usually achieved by use of a keypad, push switch, or wireless remote.

The location of the controls should be carefully planned to achieve the desired effects. To reduce use of the opener to only those needing it, the controls should be located slightly out of the general traffic path. Wall-mounted controls must be mounted so that the door to be powered is visible from the control location. Wireless wall controls simplify the installation because interconnecting wiring is not necessary.

Wall switches or touch controls require a small hole be drilled in the jamb or wall near the point where the power cord from the unit is located. Another small hole must then be drilled through the back cover of the opener to permit the low-voltage wire to be routed from the controls to the unit. Be sure to provide a strain loop on the power cord coming out of the unit so as to reduce wire fatigue from the cord flexing as the door opens and closes. Electrical hookup can now be completed. Be sure that the opener is provided with a 115-V ac grounded outlet, or hardwired in accordance with local codes.

Normally, there are only two field adjustments to be made to the unit. The first is "time hold open," and the second adjustment is for sensitivity. This safety feature is to stop the arm from opening the door when the door hits an obstruction on the opening cycle. The time-hold-open potentiometer can be set anywhere from 0 to 30 seconds. Next, set the load sensitivity adjustment to the proper position for the particular application. By turning the sensitivity adjustment to the minimum setting, the unit will be at the most sensitive setting (the door probably won't open). Turn the screw clockwise in increments until the door achieves a full-open position. During this operation the door should be manually stopped in the opening cycle, to be sure the motor stops and recycles to the closed position.

If a safety carpet is to be used, the carpet leads must be hooked to the inhibit terminals. Check its operation with the unit. If someone is on the carpet, the door will not open or close depending on the door's position. If the door is moving and someone steps on the carpet and remains on it, the door will complete the cycle and then will be inoperative until the safety mat is cleared. These inhibit terminals can also be used to interface with an elevator or lift safety circuitry.

Wiring the black and white switch terminals in parallel will allow two or more of these units to be connected for simultaneous use from one input. In the case of radio control output, one receiver can be used with the units wired in the same way for multiple uses. Note that when two or more units with radio controls are installed in the same vicinity—and are to operate independently—different codes must be set for each door system so that only the proper unit(s) will respond to a given signal.

Bifold glass and metal doors (Fig. 4.9) are normally found in commercial applications but are also being adapted for residential use. As

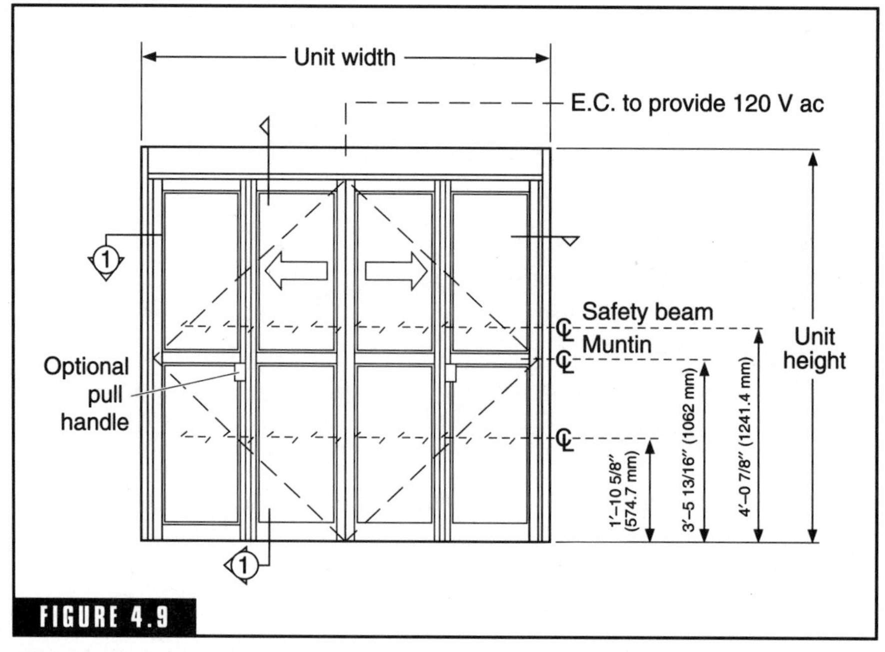

FIGURE 4.9

Bifolding doors.

an alternative to sliding patio doors or even as main entrances, these doors provide 80 percent or more of the opening as clearance. The additional clearance is useful for wheelchairs or for homeowners carrying in bags of groceries. Automating these doors, as with sliding doors, is done with low-voltage motors and rollers gliding on tracks. Control of the bifold doors is accomplished with any of the previously mentioned keypad, sensor, or wireless remote systems.

Many aspects of doors and windows relate to security issues. This topic is discussed at length in Chap. 8.

Notes

Notes

CHAPTER

5

Plumbing

The plumbing system of a home offers a variety of opportunities to use home automation. Comfort, convenience, safety, and economy are some of the prime reasons that homeowners are looking for homes using automated controls within the plumbing system.

Many homes have two types of plumbing—gas and water. Gas plumbing is primarily provided for HVAC equipment, stoves, water heaters, clothes dryers, and barbecues. These devices can be controlled through the electrical connection they also require, so additional plumbing controls for gas plumbing are seldom necessary. However, the mechanical components of gas plumbing are undergoing a few changes in new homes. New flexible tubing and quick-connect outlets (Fig. 5.1) are beginning to replace the older black-wall piping, fittings, and plugs.

One of the most significant changes involves the use of an electrically controlled shutoff valve where the gas service enters the home. Other members of a home network can also control this valve. A security system, for example, can be programmed to automatically turn the gas off when fire is detected or when the homeowners are on vacation. A PC controller or telephone responder could also be used to turn it back on as the owner is returning from vacation.

As with the gas plumbing, an electric solenoid valve can be attached to the incoming water line and used to shut off the water by

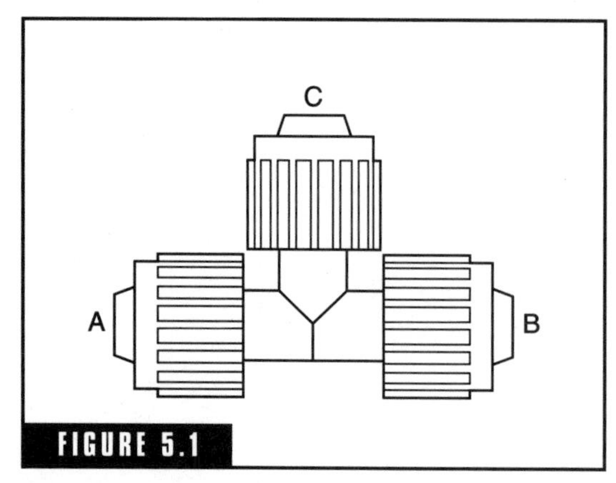

FIGURE 5.1

A quick-connect outlet.

remote control. This feature allows the homeowner to turn off the water system while on vacation, or a security system can automatically shut off when moisture is detected. Similar devices can be used to control lawn sprinkler systems and standard plumbing fixtures such as shower and bathtub faucets throughout the home.

Controlling Water at the Source

Electric solenoid valves are commonly used to control the flow of water at its source. Connected in-line, usually just past the meter location, these valves can function independently or can be part of the whole-house automation system. Sensors that detect abnormal moisture, such as a water leak would produce, are connected either directly to the solenoid control or to the home automation controller. When a leak is detected, whichever controller is used responds by sending current to the solenoid, closing the valve. Remember that these devices are in addition to a manual cutoff valve, not a replacement for them. Manual cutoff valves are necessary as backup to the electric valve and provide the homeowner or tradespeople control when working on the plumbing.

The leak sensors can be placed virtually anywhere. They are most commonly located where homes typically experience leaks, i.e., basement areas near water heaters, boilers, or washers. Placing sensors in laundry rooms, utility closets, bathrooms, and indoor spa areas is also a popular application.

Showers and Baths

Plumbing controls have several practical uses besides turning the system off during trouble. There are a growing number of appliances that include built-in controls for water plumbing. For example, manufacturers now offer complete prefabricated showers, tubs, spas, and other equipment, which attach to the plumbing in a home and include built-in electronic controls. These devices typically include more features than do devices that simply control water at its source. Some shower and bathtub faucets are controlled by electric solenoid valves, which use standard power (120-V alternating current in the United States). These valves open when voltage is applied and close when voltage is removed.

A standard controllable electronic appliance switch (similar to those used for lights and appliances) is used to operate the valve in this application. Homeowners open faucets and adjust the water flow rate and temperature to desired levels. The faucets are left open at these settings, and the switch is turned off to turn off the water flow. When the user turns the switch on, the water begins to flow again at the preset volume and temperature; in the off position, the switch stops the water flow. The switch allows the user to manually control the water flow by pressing the button on the switch face; or the user can control the water by any of the remote controls or automation controllers in the home network. Other operating systems include designs incorporating local control keypads that are attached to control valves located near the water heater. Others use specialized keypads, which include such features as individualized temperature controls.

A ground-fault circuit interrupter (GFCI) is attached to the circuit and installed near the solenoid valves. The switch itself is normally placed in a location convenient to the tub or shower. The solenoid valves and the GFCI are usually installed in a cabinet which is located near the tub or shower, and they are designed to be accessed for service. The installer can fabricate a cabinet, or some manufacturers offer a prefabricated cabinet for this equipment.

In locations subject to regular power failures, a consideration for the installer is the fact that the solenoid valves require power to turn on the water. They will not function when the power is off. To cope with this situation, the installer can add plumbing which includes a manually controlled valve bypassing the solenoid valve. This manual valve would then allow the user to turn the water on (and off) when the electric service to the house is out.

Homeowners with young children or elderly parents living in the home may need an addition to this system, or they may enjoy it just for the convenience. Installers can add a sensor that detects water level in tubs. Connecting the sensor directly to the control valve circuit will result in the water being turned off by disconnecting the solenoid when the water reaches a certain level.

Control

When these types of shower and bath valves are connected to interfaces, they provide a method to turn on the shower, tub, or spa by remote controls or in response to an automation controller. They can be used to turn on both the equipment and a heater or other device. As part of a home network, these systems allow the homeowner to utilize different methods of control. In conjunction with an automation controller, the homeowner can preset the shower or tub controls to automatically turn on when the alarm clock goes off in the morning. However, if the tub drain is not equipped with a solenoid of its own, the drain (for baths) will have to be closed prior to each filling. In the event that a water sensor is not used to control the filling of tubs, an automation controller can assist with these controls by turning off the water after a predetermined time has elapsed.

In an advanced system, other devices in the home network provide options for controlling the shower and bathtub plumbing. For example, for those of us who enjoy hitting the snooze button on the alarm clock, an automation controller can be programmed to turn on the shower or tub only when the security system detects motion in the bathroom during the hour or so after the alarm clock rings. Thus the water does not automatically start when the alarm goes off, but it does automatically start once the user arises and enters the bath or shower area.

Remote controls allow homeowners to utilize the home network to start the shower or bath from other rooms. This function can also be accessed while the person is still at the office, gym, mall, or school. Car phones can be used just as effectively as corded or wireless phones.

INTERFACING

The most common built-in controls include those used for turning on a heater or whirlpool pump or for adjusting the spray jets in showers. These functions can be remotely controlled and automated via an interface device.

Interfaces normally connect to power line wiring in order to receive control signals from the home network controller. The interface then attaches to the low-voltage circuits of the shower, tub, or spa controller. Many manufacturers have yet to include a simple connection point for interfaces; however, the installer can add the interface circuit by connecting it in parallel with the user buttons which are normally included on or near the equipment.

Automatic Faucets

Automatic faucets are actually single-temperature water control units. These no-touch faucets utilize a sensor mounted in the wall cavity or on the surface of the wall. Passing a hand or any object within 2 to 4 in over the sensor will trigger the flow of water. The flow of water may be stopped at any time by passing a hand or object over the sensor again.

As seen in Fig. 5.2, a solenoid valve from an ac or 12-V dc power pack controls the water flow. The length of time the water runs is programmed into the system at the time of installation. Most of these systems can be connected to any standard spout or outlet including thermostatic mixing valves.

The single-temperature basin represents another form of the automatic faucet or bench-mounted tap, shown in Fig. 5.3. It has a waterproof sensor into which a spout is fixed. A solenoid valve from an ac or 12-V dc power pack or from an optional battery pack controls the water flow. A control box for the unit is installed separate from the sensor in a location where there is no risk of water damage. Most of these units come complete with all components necessary for installation, including an in-line filter, which is a must-be-installed item.

Aside from the convenience of these systems, there are definite ecological and cost-saving factors involved. For example, activating the sensor directly under the outlet of the spout will cause water to flow for a predetermined time (usually 20 to 30 seconds). Should the homeowner remove his or her hands from the sensor's range while soaping, the water flow will stop. The water will only restart when the sensor is again activated. To clean the basin without activating the water flow, quickly wipe past the sensor range with a cleaning cloth. This type of operation results in cutting water use by up to 90 percent.

Automatic faucets are for bathtub applications, as seen in Fig. 5.4. These systems usually provide a choice of temperatures and time cycles so individual family members can set their own comfort levels. A manifold with solenoid valves completely controls the water flow.

To operate the system, hold a solid object (e.g., a hand) within 2 to 4 in in front of the sensor, and a display will come on. The display

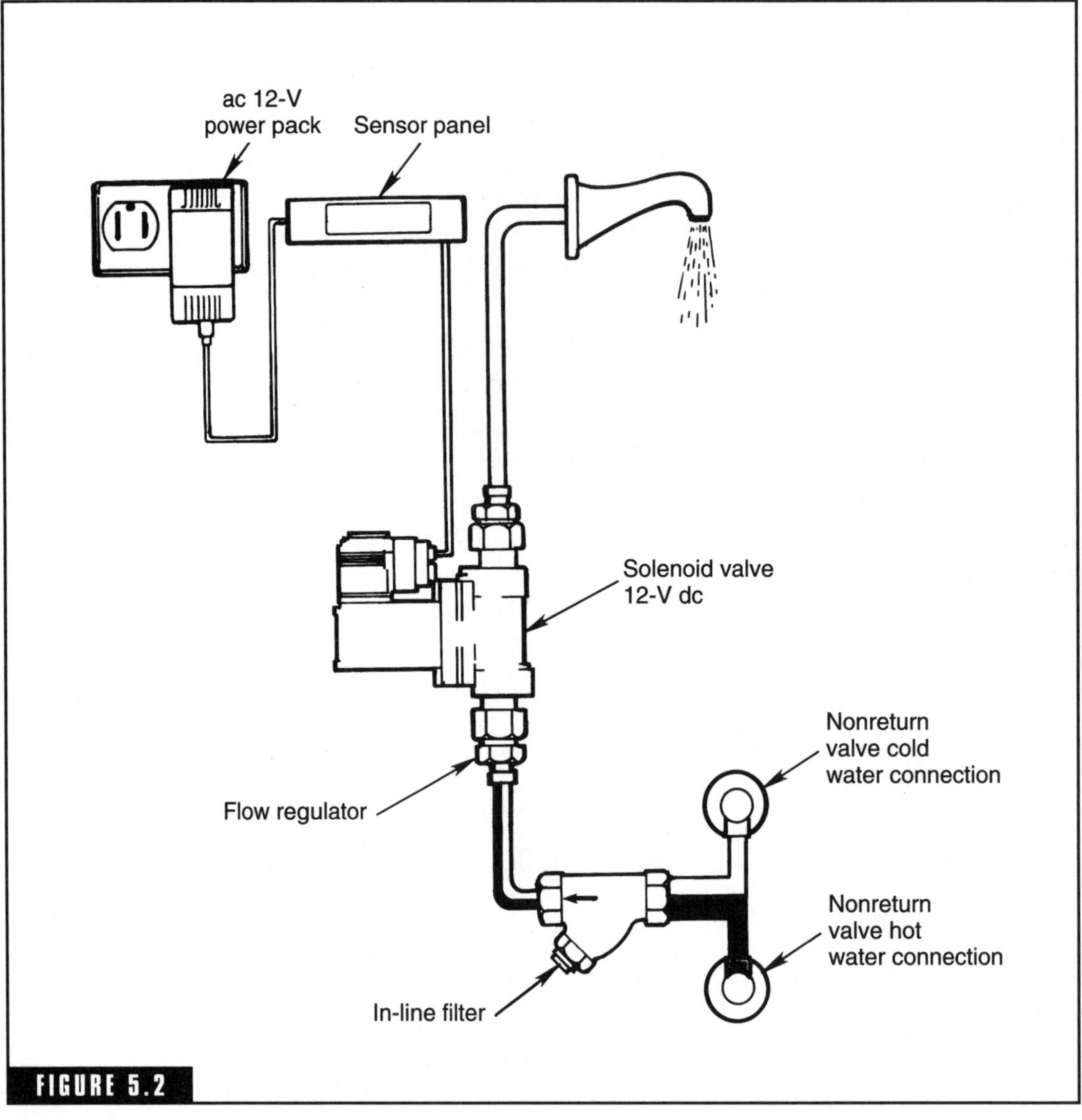

FIGURE 5.2

No-touch faucets utilize a sensor mounted in the wall cavity or on the surface of the wall.

will indicate the length of time that the faucet is to run and the temperature of the water. The display will cycle through a sequence, once the desired time and temperature are displayed; removing the object from in front of the sensor will start the water flow. The water will flow at the desired temperature for the time selected. To stop the water flow before the end of the preset time cycle, pass a hand over the sensor.

Automatic Toilets

Most of us, when visiting rest rooms in sports arenas, industrial sites, or commercial locations, have encountered toilets that flush themselves. This technology is now becoming popular in high-end residential buildings.

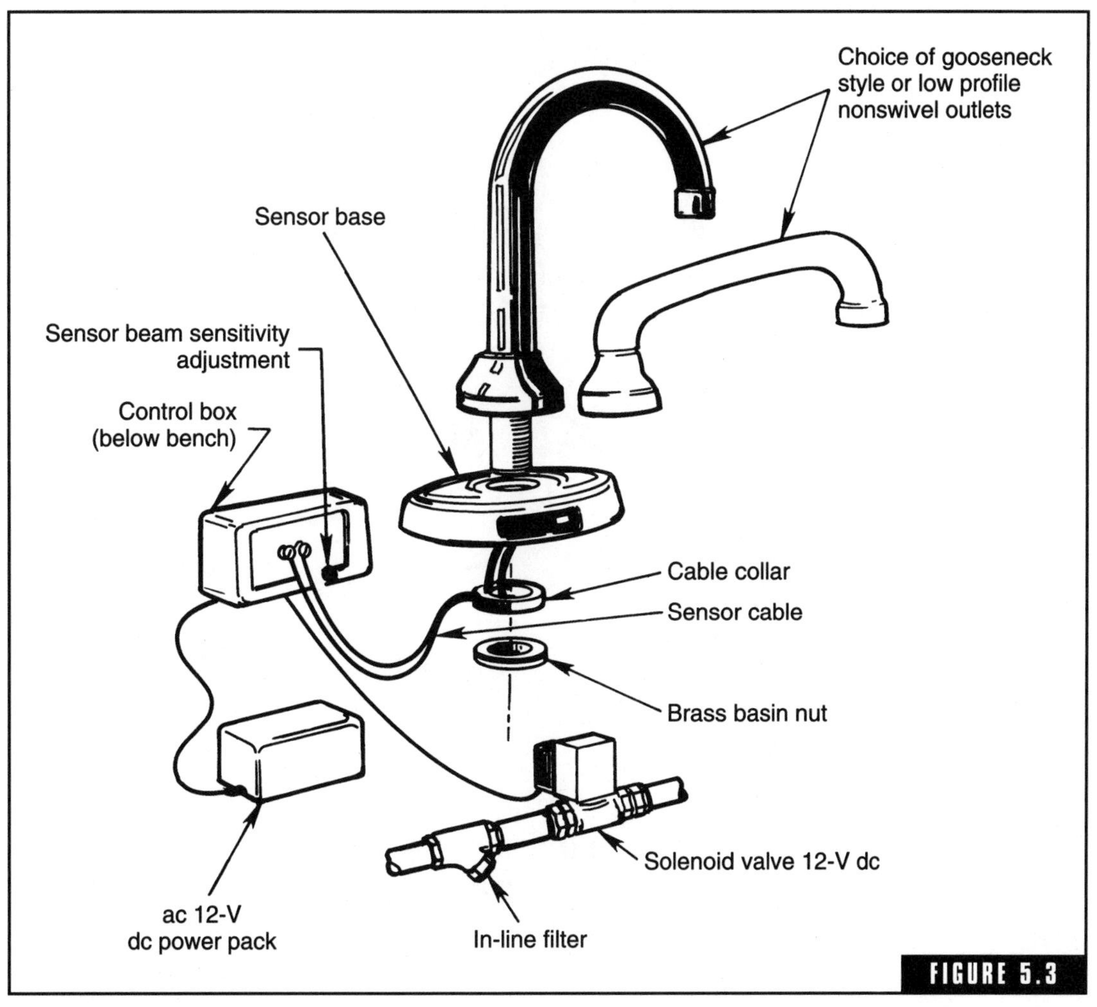

A single-temperature basin.

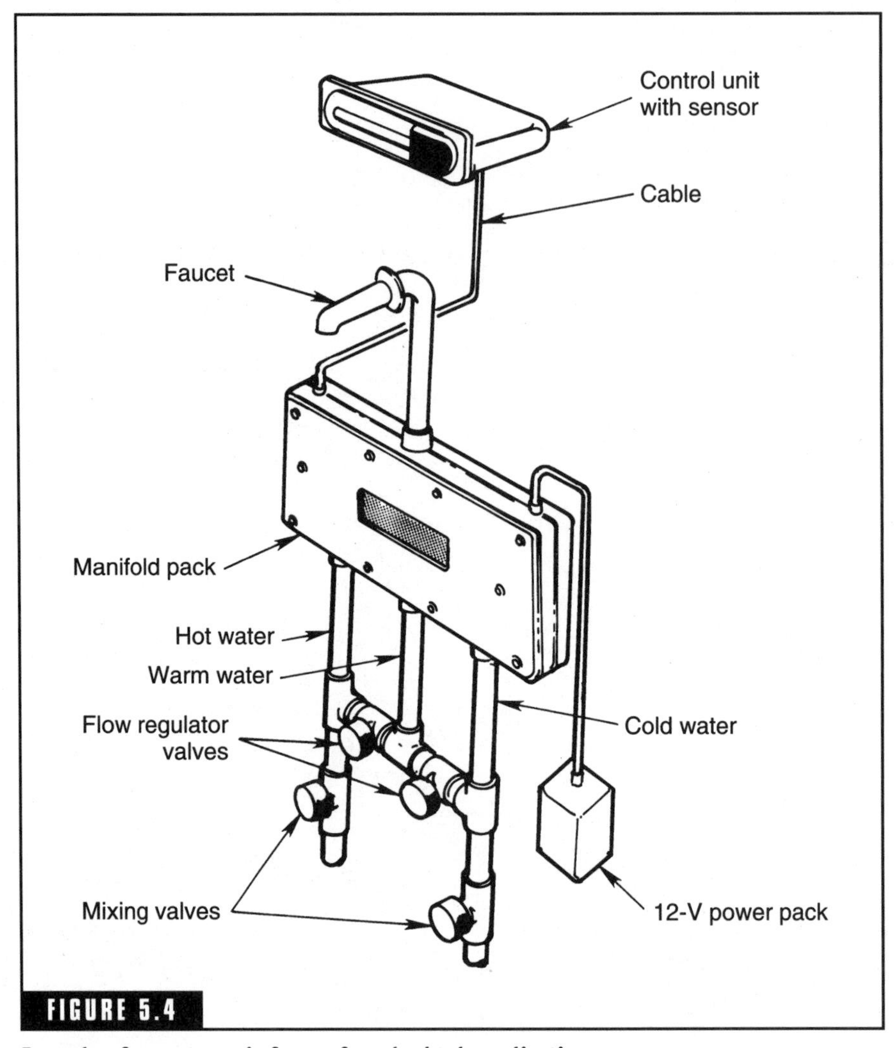

FIGURE 5.4

Example of an automatic faucet for a bathtub application.

These toilets, like the automatic faucets, have sensors, which activate the mechanism. These sensors, as seen in Fig. 5.5, are usually either wall- or ceiling-mounted and are linked to a multiple-choice control box. The water flow is again controlled by a solenoid valve from an ac/12-V dc power pack, or by an optional battery pack. When the toilet sensor perceives a presence, the controller places the toilet in a preflush mode. When the sensor perceives the absence of a presence, the controller performs a full flush.

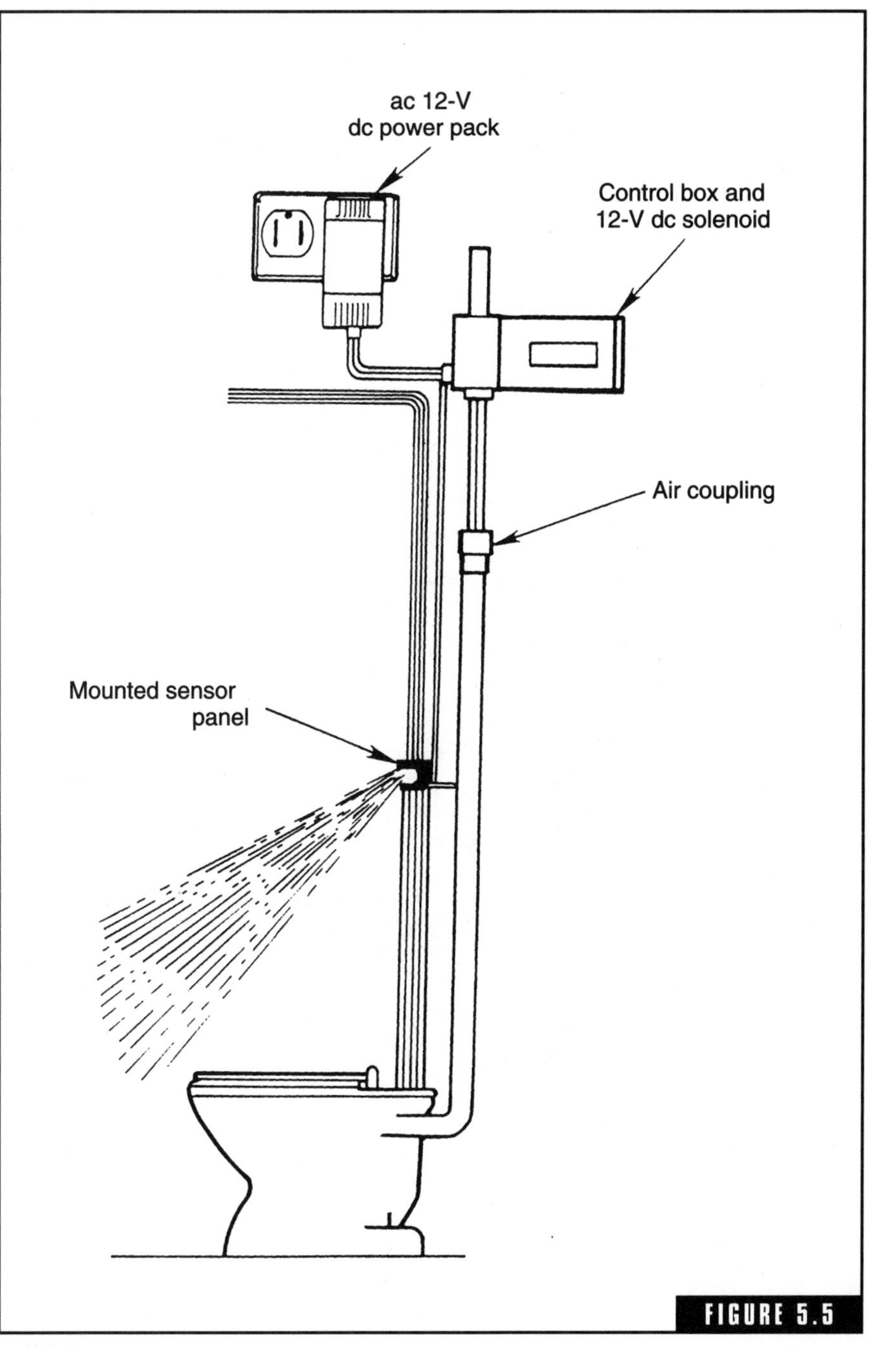

Automatic flush mechanism.

Temperature Balancing

Most people have had the unpleasant experience of a change in water pressure while taking a shower. Many times the results are either a freezing or scalding interruption in the water temperature. Pressure-balancing maintains the selected shower temperature by instantly and continuously balancing pressures of the hot and cold water systems, compensating for changes in pressure made by other water users. An example of how these devices operate is demonstrated in Figs. 5.6 and 5.7.

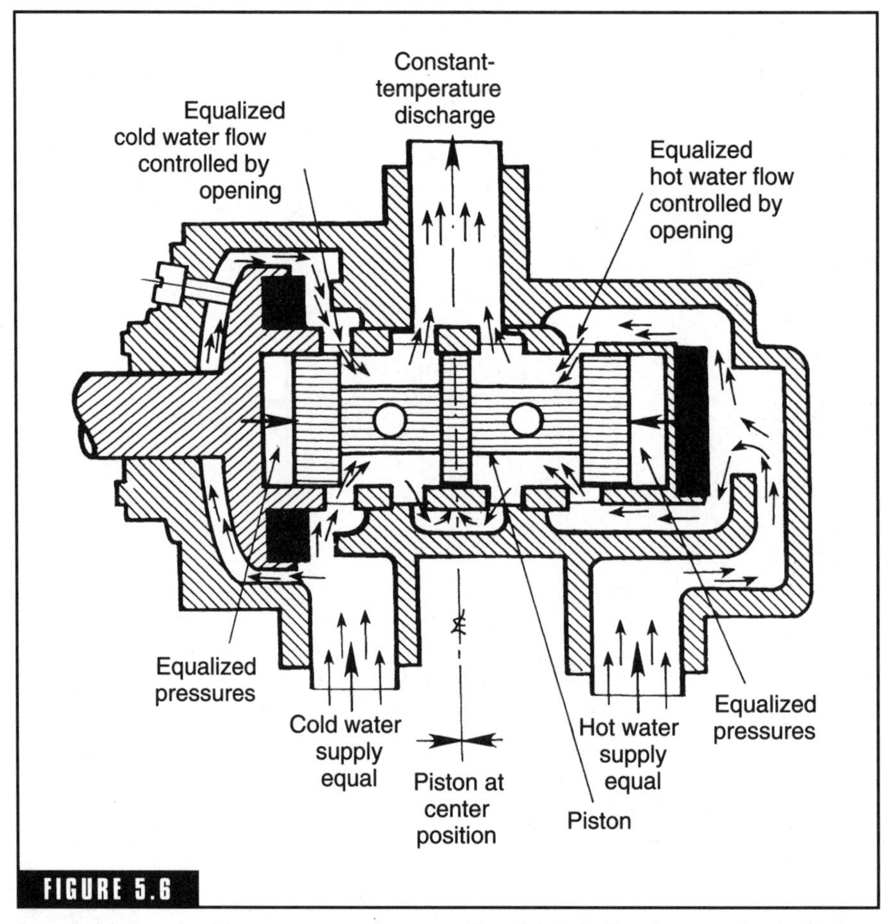

FIGURE 5.6

An equal hot and cold pressure supply moves the piston in the shower valve to its center position, allowing equal flow of both hot and cold water.

An equal hot and cold pressure supply moves the piston in the shower valve to its center position, allowing equal flow of both hot and cold water. When pressure in the cold water supply drops, the piston moves to the cold end, reducing the opening for the hot water supply. Thus the temperature remains constant (Fig. 5.7).

When pressure in the hot water supply drops, the piston moves toward the hot end, reducing the opening for the cold water supply (Fig. 5.8). Again, temperature remains constant. In the event of a cold water supply failure, the piston moves to the extreme position on the

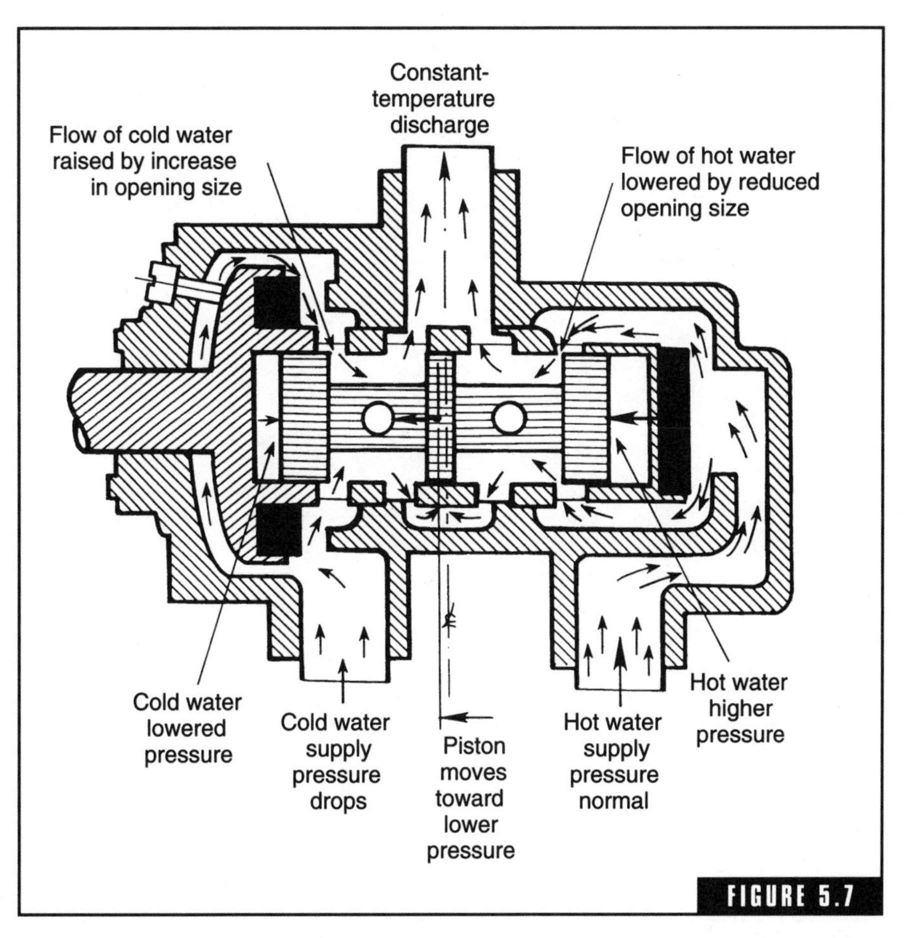

FIGURE 5.7

When pressure in the cold water supply drops, the piston moves to the cold end, reducing the opening for the hot water supply.

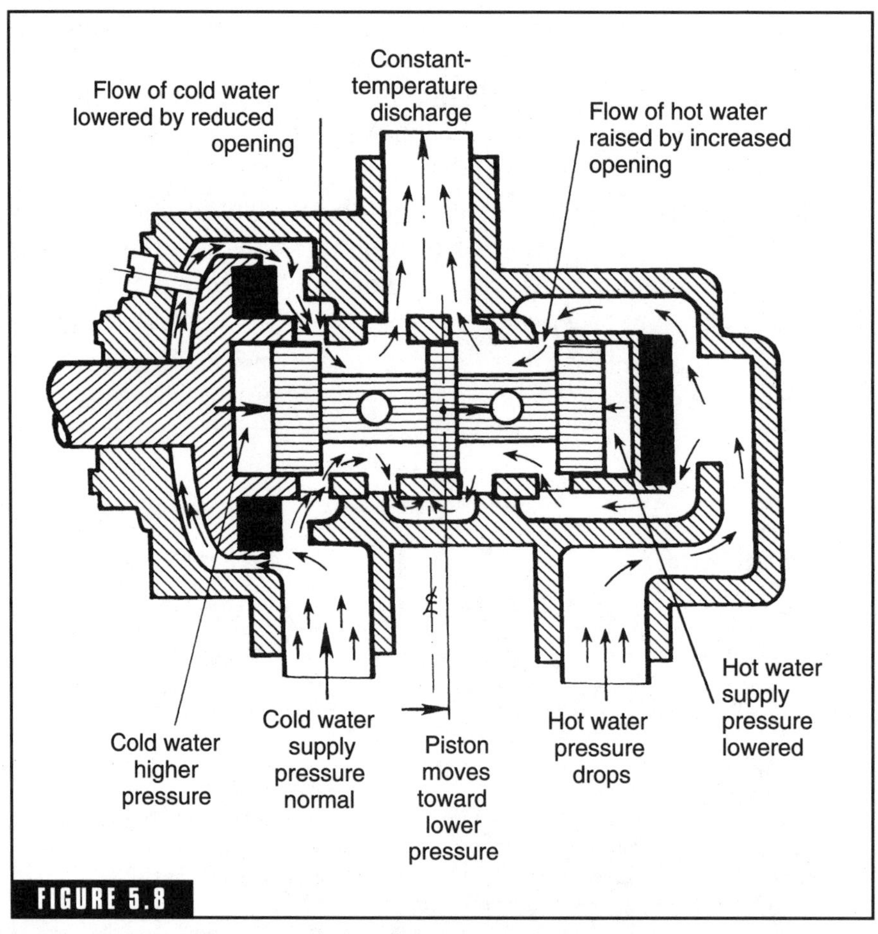

FIGURE 5.8

When pressure in the hot water supply drops, the piston moves toward the hot end, reducing the opening for the cold water supply.

cold end, which cuts the hot water supply completely (Fig. 5.9). All water flow through the valve is shut off.

Water Softeners

Water is referred to as the *universal solvent.* Given enough time, water will dissolve any organic material. It surrounds foreign particles, such as minerals, entrapping them in what scientists refer to as *complexes.* That's why water usually has a high mineral content.

These dissolved minerals are not a part of the water itself. They are captives that the water has surrounded and is carrying along with it. The number of mineral complexes in water determines how "hard" the water is. The more minerals it carries, the harder it is considered to be.

When water is stored, is heated, or evaporates, the complexes carried are broken up and the dissolved minerals are set free. These liberated minerals (most of which is calcium carbonate or magnesium) form sediments that line the insides of pipes, appliances, water heaters, and other surfaces with which the water comes in contact. Over time, more and more minerals build up on the sediment layer,

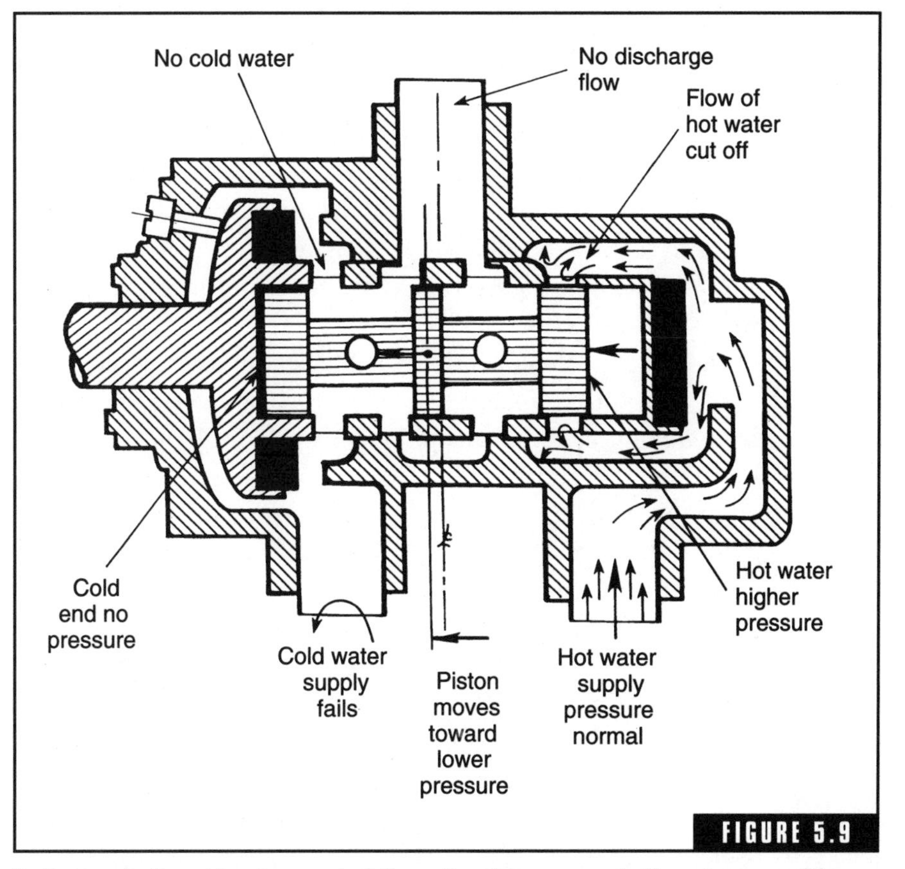

FIGURE 5.9

In the event of a cold water supply failure, the piston moves to the extreme position on the cold end, which cuts the hot water supply completely.

causing it to grow progressively thicker. There is a name for these caked-on mineral deposits: *scale.*

Conventional water softeners, which filter water coming into the house through salts contained in a tank, take some of the hardness minerals out of the water and replace them with sodium. These water softeners are pretty much automatic. The hands-on aspect of these devices involves the salts. The cost of continually replenishing salts is one of the drawbacks to conventional water softeners; some people also notice a change in the taste of the water.

Magnetic Fluid Conditioners

One alternative to conventional water softeners is a *magnetic fluid conditioner.* Instead of using salts to filter hard water, these units are powered by highly intensified, computer-designed, magnetic fields requiring no outside power source. *Magnetohydrodynamics* is the scientific name for what occurs when water passes through a properly focused magnetic field. Some of the complexes that are carried in the water are broken up by the magnetic field, freeing the captive mineral particles. Once free, these particles give the surrounding mineral molecules in the water something to stick to, rather than forming layers of scale on the inside of plumbing appliance surfaces.

Minerals in the form of suspended particles, contained in virtually all fluids, are physically affected by exposure to powerful magnetic fields. After treatment each particle adopts a surface charge. As these particles flow along the plumbing surfaces, contacting more deposited material, a charge transfer occurs. This charge transfer causes a disruption in the state of any deposited materials. Deposits are slowly eroded away by the scouring action of the fluid flow.

Magnetic fluid conditioners must be installed on nonferrous materials (copper, plastic, PVC, rubber, aluminum, stainless steel) to function properly, as seen in Fig. 5.10. The penetration of a magnetic field through steel tubing or pipe (except for stainless steel) or iron pipe is minimal. If the magnets stick to the tubing or pipe, they won't work. The accepted method of treatment in this case is to remove and replace an 18- to 24-in section with copper, plastic, PVC, rubber, aluminum, or stainless-steel tubing or pipe.

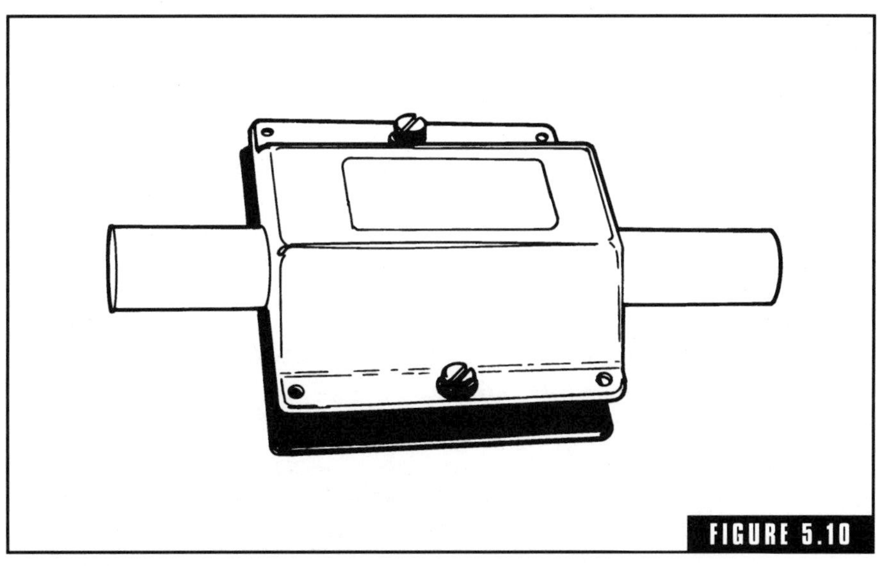

FIGURE 5.10

Example of a magnetic fluid conditioner.

For best results, always install these units a minimum of 24 in downstream of any pump, elbow, or other turbulence-causing device. Ideally, magnetic treatment requires a laminar (smooth) flow of the fluid through the magnetic field for maximum effectiveness. In addition, the interaction between the energy of the turbulence and the treated fluid may cause a partial reversal of the treatment.

The magnetic treatment of the system may also be decreased if electric equipment capable of generating a strong electric field, i.e., a welder or large electric motor, is too close to a water line on which the units are placed. Similarly, an electric power line running parallel to and within 6 in of the pipe may decrease the effectiveness of the treatment. Electric lines crossing a pipe have no effect.

WATER CLARITY

When installed in existing homes, these devices may cause an increase in water turbidity (cloudiness). This relates to a stabilizing period of a few weeks during which calcium deposits are being removed from pipes and appliances while the system is descaling. During this period it is important to monitor filters, tanks, and sumps weekly and to drain or clean them as needed.

Water Heaters

An appliance that directly benefits from reduced mineral content in the water is the water heater. Water heaters are also one of the biggest consumers of energy in the home, running up utility bills even when water consumption is zero. To control the operation of a water heater requires that a switch be installed to facilitate remote control. With an electric water heater a controlled appliance switch can be placed in the circuit powering the water heater. Other options include controlled receptacles and plug-in modules. However, the use of appliance switches provides the additional benefit of a manual switch for local control.

For water heaters using gas instead of electricity, an electrically controlled solenoid valve can be placed in the line to the main burner. The solenoid is then connected to an interface that enables control of the device through the power line.

Manufacturers of water heaters are designing new units to be more energy-efficient. One of the design changes has been to include more insulation for the tank. This additional insulation reduces heat loss through the tank walls, retaining the heat in the water for greater periods. Turning off the water heater when it is not in use can further reduce utility bills. Using this type of control to limit the number of hours per day that the unit runs will also increase the life span of the product.

Once the installation of a remote-controlled switch is accomplished, an automation controller can be used to turn the water heater on and off at predetermined times. These times will depend on the lifestyle and preferences of the homeowners. Installers should cover hot water needs as part of their preinstallation interview with the homeowners; getting the proper setting the first time will eliminate costly callbacks in the future. For example, a family that bathes or showers in the morning should have the water heater turned on about 30 minutes to 1 hour prior to the time the first hot water demand occurs. If the house is to be unoccupied during the day, the hot water heater can be turned off until later in the afternoon when the family returns. Evening schedules should include hot water demands for dishwashers, clothes washers, and additional bathing requirements placed on the system. Once the evening demands for hot water have been met, the water heater can be turned off until the following morning.

Water heaters can also be tied into the home's security system to provide yet another method of control. An interface within the security system that allows control signals to be sent throughout the home network provides some very useful alternatives in controlling the water heater. The system can be programmed so that when the security keypads are used by the homeowner when leaving or arriving, the water heater will automatically be turned off or on, respectively. Along the same lines, the security system can utilize motion sensors to turn on the water heater when motion is sensed. Conversely, the water heater automatically turns off when everyone leaves the house or goes to bed. As with other applications, the security system or automation controller can be programmed for vacations, leaving the water heater off indefinitely while the homeowners are away.

Pools and Spas

More and more frequently pools and spas are incorporating automation into their installations. Intelligent controls that automatically keep the water level at preset limits or regulate water temperature as well as detect when a person enters the water are popular in high-end homes.

A solenoid valve and moisture sensor that are connected to a plumbing source easily take care of the automatic filling functions. An independent device using one or more sensors can be added to the pool or spa to detect when it is entered. Typically connected to the security system, these sensors are both popular and practical for families with small children. Also many homeowners find this type of warning system useful in today's litigious society. Uninvited guests or strangers falling into pools or spas (even if they did jump over the locked fence) are suing and winning in court. A simple sensor system can help homeowners protect themselves from unexpected problems.

The best known and most common automation controls used for pools and spas involve lighting controls, cycling pumps, and regulating heaters. Pool lights can be designed with switches that can manually turn lights on and off while allowing the automation network to take control as well. Lights can be set for dusk-to-dawn operation or can utilize motion detectors so they are on as long as there is motion present within the sensor's range. Other lighting control can be achieved by using the security features of the network. For example, the security system can be programmed with the normal, seasonal times when it is dark outside. With doors or gates leading to the pool or spa area wired into the security system, lights can be automatically turned on when one of these doors or gates is opened after dark.

Pool and spa heaters, by design, are already automatically controlled since they only function when the filtration or circulation pump is on. Pumps are usually hardwired into the home's electrical system; occasionally a spa pump will be designed to be plugged into a 220-V receptacle. Automating these pumps is easily done with the use of electronic appliance switches or a controlled receptacle. When the switches or receptacles are installed, homeowners can control these devices remotely, turning them on or off with remote controls or an automation controller. Again, it is important for the installer to be familiar with the homeowner's lifestyle. Running the pump for a few hours in the middle of a sunny afternoon can take advantage of the sun's energy to warm the water. On cloudy or rainy days, air temperatures may be too cool for swimming. On these days the automation system can be programmed to run the pump for a minimum time to save energy. The *scene* feature of home automation controllers can be used to create a predefined atmosphere for entertaining. Part of a particular scene may include the pool or spa area. In this scenario, for example, the network could automatically turn on the pool lights and patio lighting and start the pump and heater at the same time as the gas grill is lighted.

Many spas include their own timers and control panels. An interface added to the circuits and parallel to the control panel wiring allows the homeowners to turn the spa on or off by remote control. Homeowners can activate the spa by using remote control codes from phones at the office or mobile phones from the car on the way home.

Laundry Areas

In many homes, particularly the larger, high-end homes, laundry areas are intentionally set apart from primary living space. Remote location makes sense; it isolates the occupants and their guests from the noise of washers and dryers as well as the additional heat and humidity given off by the machinery.

On the downside, remote laundry areas have presented a dilemma for ages: When is it time to switch loads? Many washers and dryers do not come equipped with buzzers to signal the end of a cycle or the end of the run. Machines that do have buzzers become less effective when placed in remote locations. Fortunately, home automation components can be adapted to rectify this situation.

Clothes Washers

As an example of how home automation can take the guesswork out of laundry day, Fig. 5.11 illustrates a simple solution. Mounted inside the control panel of most washers is a clock with a 120-V ac motor. Run the motor in parallel with a 120-V ac relay, and run a pair of wires from the common NC contacts to the input of an X-10 PowerFlash module. Plug an X-10 remote chime module into a standard wall outlet located in an area convenient for the occupants. Now, when the

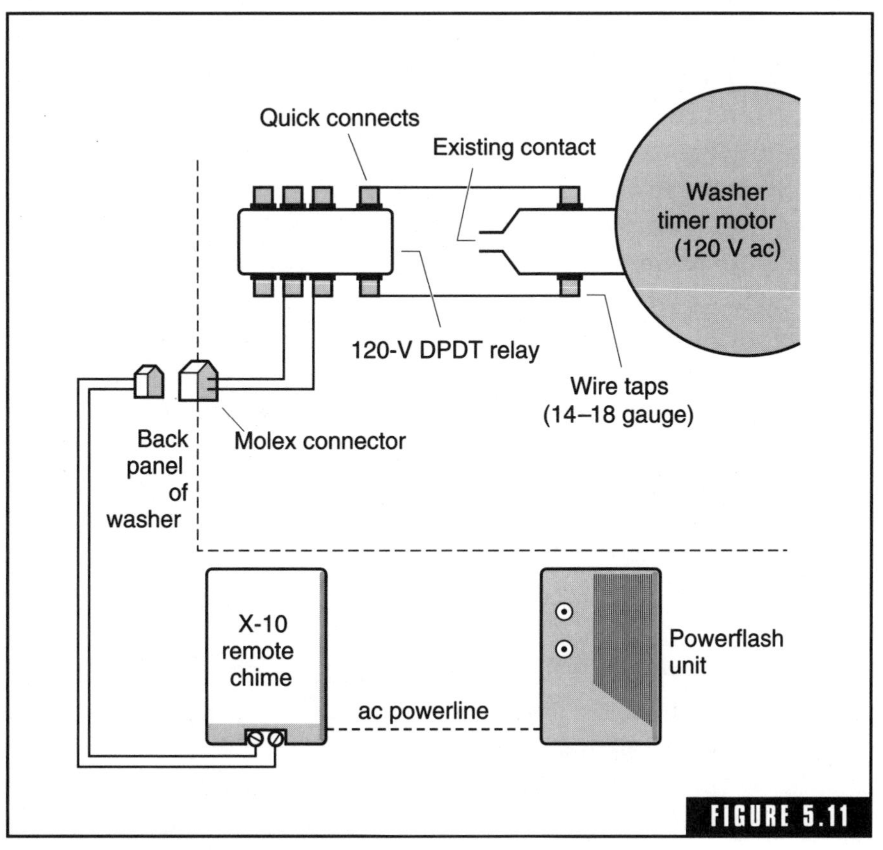

FIGURE 5.11

Adding home automation to the washing machine.

washer finishes spinning, the PowerFlash module will send signals over the power line to the remote chimes, alerting residents that it's time to transfer the load to the dryer.

The different cycles of a washing machine are controlled by the interaction between a timer motor and a water-level switch. A timer motor is a complex timing scheme that involves multiple contacts triggered by a single clock. The timer motor runs in conjunction with the washer, except when the water level drops below required levels for the next cycle. To increase the water level enough to begin the next cycle, the water level switch opens; during this *fill cycle,* the timer motor's function is suspended. The water-level switch closes after the first spin cycle is completed, and the timer motor resumes running until the final spin cycle is completed. Because of this interaction, no single timing position can cause a buzzer to sound.

Installing the relay by connecting it across the timer motor circuit will result in the NC contacts opening when the washer is started, which pulls the relay in. After the first spin (when the washer pauses), the relay is activated, closing the contacts. This action causes the PowerFlash module to send a signal to the chime module, sounding the first notice. When the motor starts up again, the relay contacts again open until the final spin cycle is completed. At that time, the contacts close, sending another X-10 signal to the chime module, signaling that all cycles have been completed.

The PowerFlash module has an NO (normally open) input. A short circuit will trigger the module to send an X-10 signal. If the short circuit is maintained, the PowerFlash module will send only one signal, then stop. Therefore, the relay is wired so that the contacts open when the timer motor is on and close when the motor is off.

Clothes Dryers

Some clothes dryers are equipped with buzzers, as washing machines are; others are not. However, with remotely located laundry areas, even those with buzzers are difficult to hear in various parts of a larger home. As with the washing machines, home automation can supply a remedy.

Since various makes and models of dryers operate differently, one definitive solution cannot be put forth. However, the following example demonstrates how home automation can be adapted with dryers as it is with washers.

A PowerFlash module can be triggered to send an X-10 signal when current is drawn, by using an ac sensor. The current sensor (Fig. 5.12) has to have a hot wire passed through it. A point where the incoming power cable connects with the dryer can be found on the inside of the dryer housing. Identify the hot wire at this location. If this wire is long enough, it can be disconnected and passed through the sensor. Loop the hot wire, passing it through the sensor twice before reconnecting it. If the wire is not long enough, splice an additional wire into the line as needed, making sure to adequately insulate all connections (Fig. 5.13). Connect the two leads on the sensor to an X-10 PowerFlash module. The PowerFlash module should be set to input A, mode 3 (on/off) and connected to an ac outlet.

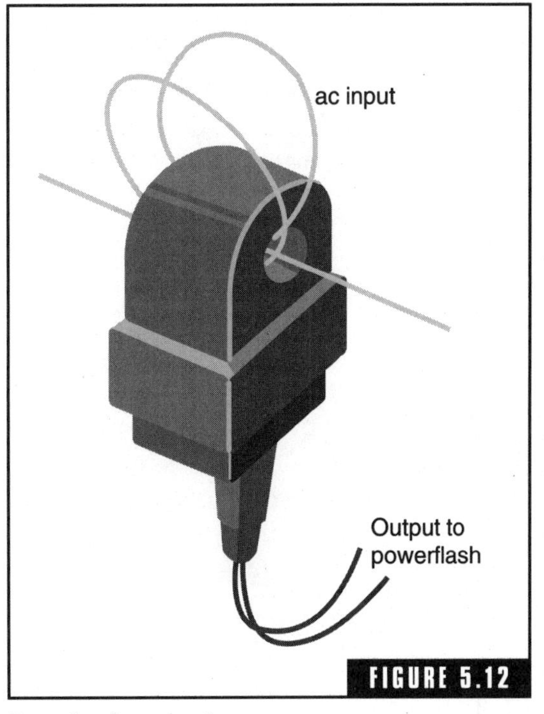

FIGURE 5.12

Example of a current sensor.

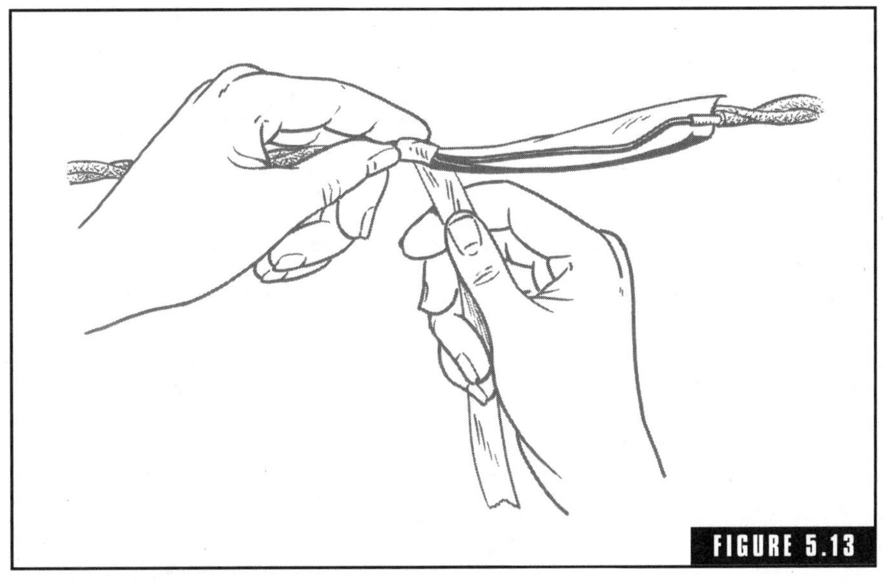

FIGURE 5.13

Insulating connections.

When the dryer is operating, the current sensor will generate a small (6- to 7-V) alternating current. This will cause the PowerFlash to send an on signal to remote X-10 devices, such as a light or a chime module, that are set to the same house code and unit code as the PowerFlash. When the dryer stops, so does the current being generated by the sensor. This lack of current causes the PowerFlash to send an OFF signal to the X-10 modules, set to the same house and unit code.

This example uses the simplest automation option—turning an X-10 module on or off. By tying the dryer into an automation controller, many more options become available. For example, the output of the PowerFlash can be used as a trigger for the automation controller to place a telephone call. The controller can call the homeowner at a preselected location (neighbor's house, pager, or cell phone) to deliver a voice message that the dryer has stopped.

More ancillary uses for home automation are covered in detail in Chap. 10.

Notes

Notes

CHAPTER 6

Communications

Telephone Lines in the Home

As lifestyles continue to change, demand for technology-friendly homes continues to escalate. More and more homeowners are commuting from the bedroom to the home office. Children are becoming more computer-literate than their parents, and schools encourage students to use home computers for assignments. The average family today can easily put more than one phone line to good use. The number of phone lines for a family will depend on how each line is to be used. A family of two adults and two teenagers may very well need two or three individual phone lines. Add a home business or home office requiring faxing capabilities and on-line data exchange, and the number of phone lines may increase to four or even five. Some of the more popular methods and technologies used to prepare homes for high-technology communications are covered in this chapter.

What Extra Telephone Lines Do for Homeowners

As mentioned previously, it is imperative that the needs of the homeowner(s) be clearly understood by builders, contractors, and installers, to ensure the proper prewiring of the home. Probably most crucial to this understanding is that the homeowner realize what options are available and what benefits can be expected from multiple phone lines. Installers, builders, architects, and contractors should all be anxious to explore the family's needs and explain the benefits of various options.

Benefits of Multiple Phone Lines

FAMILY PHONE LINES

Two or more phone lines avoid the competition for dial tone. If one family member is using one phone line, outgoing calls can still be made from a second line. Parents who have teenagers or preteens can usually appreciate the idea of a dedicated line for the children.

BUSINESS PHONE LINES

Home offices and businesses present a more professional appearance with a phone line dedicated for that use. Customers or business associates expect a business phone to be answered by their associate or voice mail, not one of the children. Plus, anytime during the day, or after business hours, the office phone can be used as a second line for personal calls.

Many home offices include two, three, or more dedicated phone lines. An individual line connected to the fax machine enables the home worker to continue making business calls while waiting for an important fax. Faxes can also be sent in a timely manner while the home worker waits for an important call. The dedicated fax line can also be tied into the computer's modem, providing a clear path for data exchange over the Internet, a local-area network (LAN), or a wide-area network (WAN).

Many people prefer to have separate lines for both fax and modem. The reason is that if the modem is in use, incoming faxes cannot get through. Three business lines separate from the personal lines in the house are not uncommon. Three business lines provide access for the modem without tying up the voice or fax line. Faxes can be sent or received while the homeowner makes voice calls and/or is participating in a computer network meeting, all without interfering with each other.

Topology

The routing and connecting of cables is referred to as *topology.* Each different arrangement of wiring comprises a different topology, and each topology has advantages and disadvantages. The topology in new construction is usually of one type. However, remodeling homes or building additions to existing homes often require the combining of topologies. The installer's primary goal is to make sure that all the jacks have continuity to the *network-interface device* (*NID*) after the wiring is done. The NID is the device installed by the phone company that connects the house wiring to the telephone network.

Star Topology (a.k.a. Home Run)

Star topology runs separate lines to each jack in the building from a central hub. Generally the hub of a star topology is located near the center of the structure in an attic, basement, garage, or other well-protected but accessible area. As seen in Fig. 6.1, individual lines leave the hub and fan out to the various jack locations throughout the house.

In some cases, if the design calls for only a few jacks, the NID itself can be used as the central hub of the star. However, using the NID as

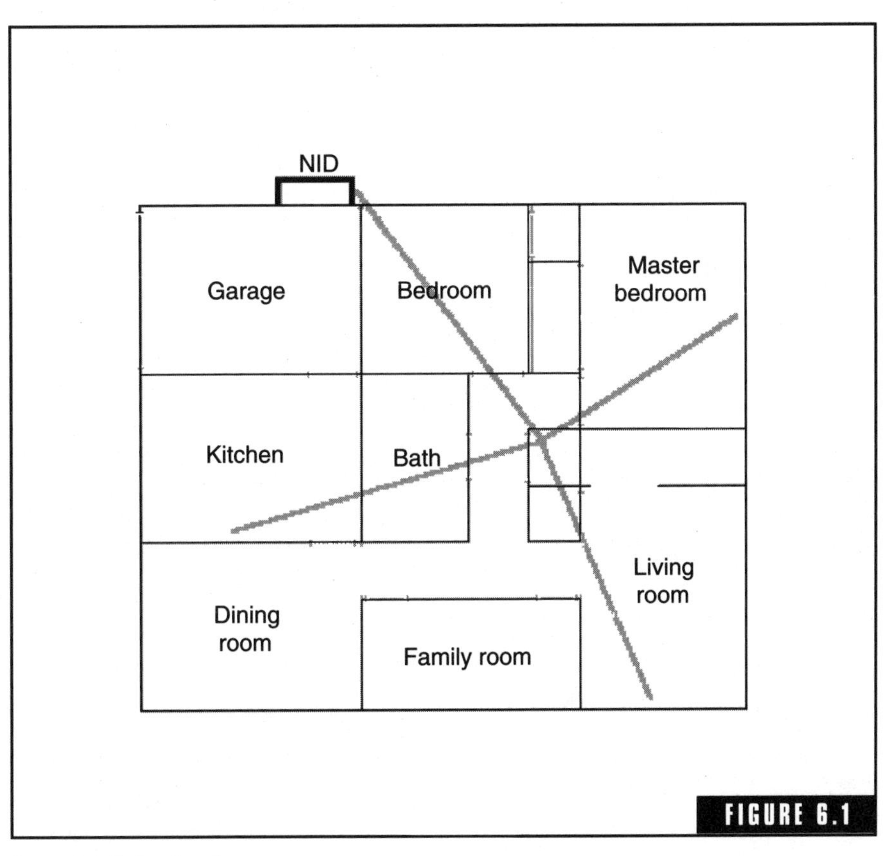

FIGURE 6.1

Example of star topology.

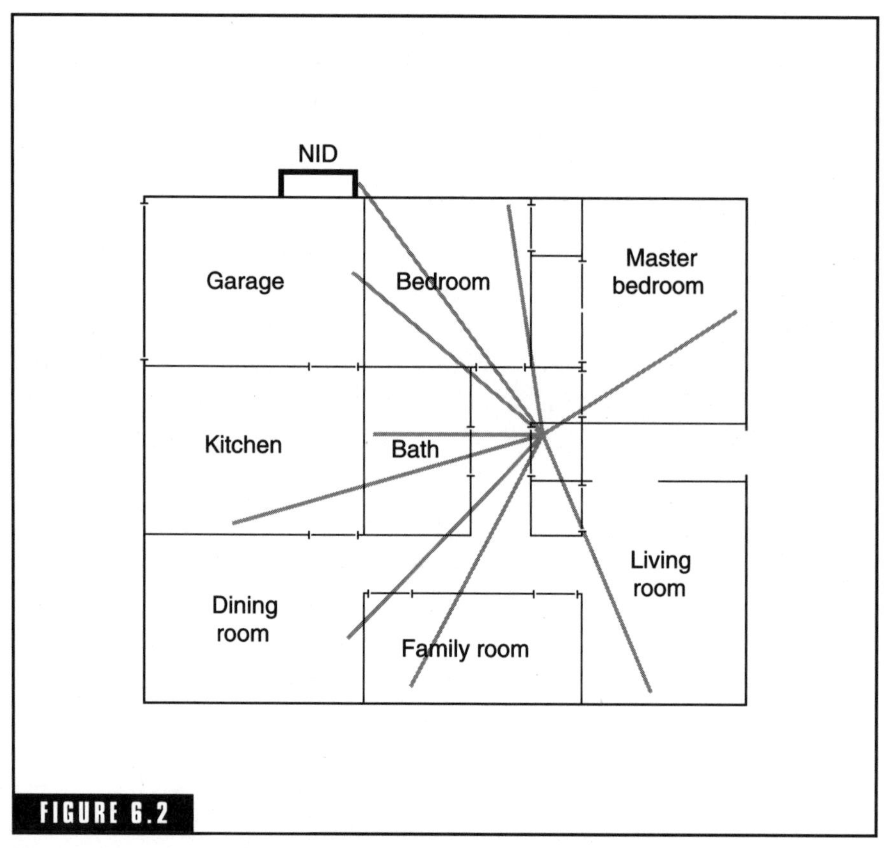

FIGURE 6.2

Star topology with additional lines.

the hub for more than just a few jacks can quickly become confusing and complicated. Installations requiring multiple jacks (Fig. 6.2) are best served by employing a junction block as the central hub. A separate cable running from the box to the NID connects the two.

ADVANTAGES OF STAR TOPOLOGY

- Additional jacks can be wired whenever the need arises.
- Broken, cut, or otherwise damaged wire going to one jack will not normally cause problems at other jack locations.
- The star configuration makes it easy to add multiple applications to the system.

- Troubleshooting of wiring problems is much easier than with other topologies.
- Star topology is the easiest configuration to convert to future technology such as an Integrated Services Digital Network (ISDN).

DISADVANTAGES OF STAR TOPOLOGY

- Generally, star topology does require more cable than other topologies.
- Unless properly installed, the hub and its connections are potential failure points.

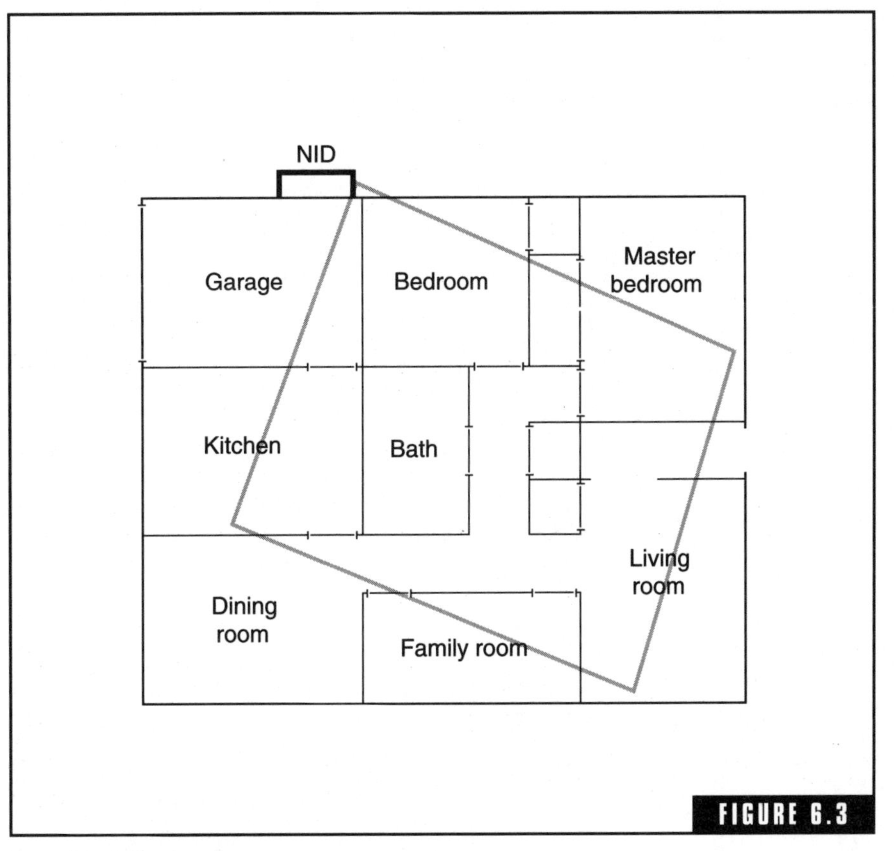

FIGURE 6.3

Example of ring topology.

Ring Topology

Jacks throughout the house are tied into one cable, forming a loop. Both ends of the cable are then connected to the NID (Fig. 6.3). To connect the wires to the jack, the sheath of the cable is removed as it comes to a jack. Some installers cut the line at each jack and connect to each jack from both sides of the cut. After each connection, the cable is run to the next jack.

ADVANTAGES OF RING TOPOLOGY

- Because jacks are actually tied to the loop from two directions, signals have two paths on which to travel to the NID. If a cut or broken wire occurs on either side, it will not normally cause a loss of service at any jack.
- The amount of cable used for the initial runs is usually less than that required for star topology installations.

DISADVANTAGES OF RING TOPOLOGY

- If not properly installed, the multiple junctions required to expand and maintain this type of topology are potential failure points.
- In most cases ring topology is more difficult to troubleshoot than star topology, should wiring problems arise.
- Additions to the system often result in reconfiguring to a hybrid type of topology.
- Initial installation tends to use more cable than bus topology does.
- Adding multiple applications to the system presents difficulties.

Bus Topology

As with ring topology, all the jacks in the house are tied into one cable that starts out from the NID. The difference, as shown in Fig. 6.4, is that the cable terminates at the last jack location instead of being connected to the NID. All the jacks in this arrangement—except the last one, where the cable terminates—are connected in the same way, as they would be for a ring topology. Existing homes that were prewired during construction most likely have lines run in this manner.

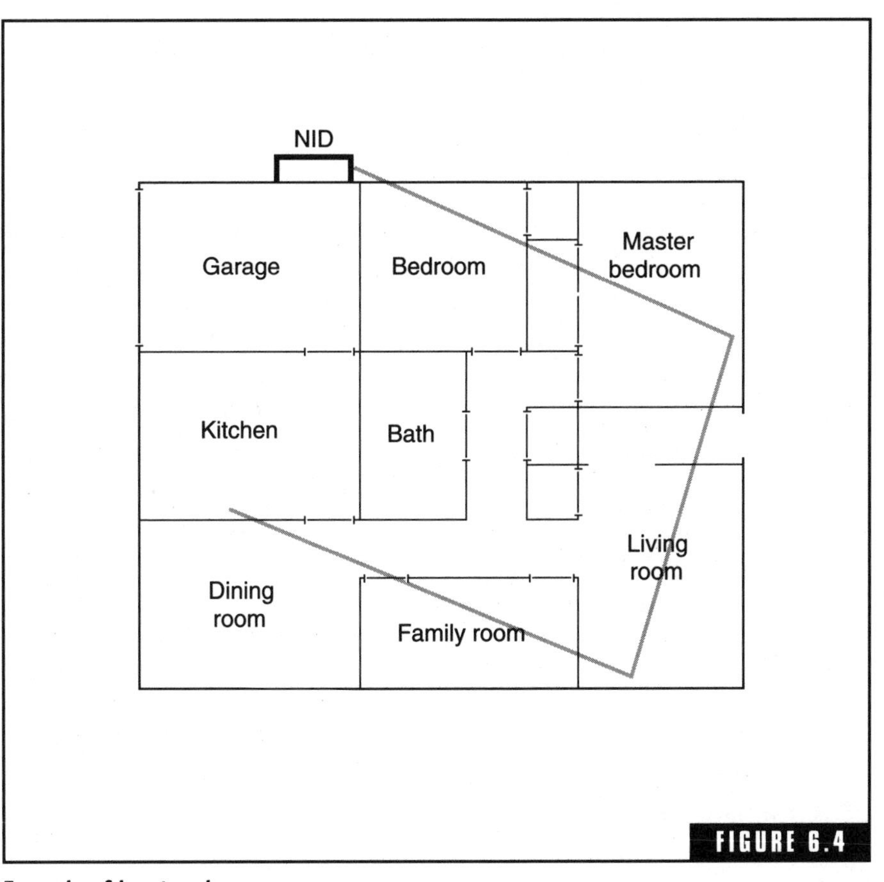

Example of bus topology.

ADVANTAGES OF BUS TOPOLOGY

- Normally, bus topology will require less cable than other topologies.
- Troubleshooting wiring problems are easy to solve.
- Additions to the system do not usually result in reconfiguring to a hybrid type of topology, although bus topology provides no significant advantage over a hybrid.

DISADVANTAGES OF BUS TOPOLOGY

- A cut or broken wire will cause loss of service at all jacks past the break.

- Adding multiple applications to the system is difficult.
- Adding to existing installation almost always results in a hybrid topology.

Hybrid Topology

Hybrid topologies normally begin life as one of the basic topologies already discussed. They become hybrid as the result of additions to the system. During the course of remodeling, building an addition, or just extending the wiring within an existing home, every effort should be made to preserve the existing topology. Star topologies should be preserved at all costs; they offer the greatest flexibility for today's technological uses and adaptability for tomorrow. If the existing lines are ring or bus configurations, there may be no other choice than to convert to a hybrid topology to accommodate additional needs.

Hybrid topologies usually consist of tap lines run off the main line in either a ring configuration (Fig. 6.5) or a bus configuration (Fig. 6.6). Changes made in any topology should be documented for future reference. A copy should be given to the homeowner and another copy kept by the installer.

ADVANTAGE OF HYBRID TOPOLOGY

hybrid topologies offer easy accommodation for additional jacks for one or two lines.

DISADVANTAGES OF HYBRID TOPOLOGY

- Troubleshooting of wiring problems is definitely more difficult than with other configurations.
- A cut or broken wire may cause a loss of service at one, some, or all of the jacks, or could result in no loss of service at all, depending on the original topology and where the break occurs.
- The necessity of using multiple junctions in a hybrid system results in potential failure points.
- Hybrid configurations are very difficult to convert to an installation of three or more lines.
- Changing over to ISDN technology or other future conversions is very difficult.

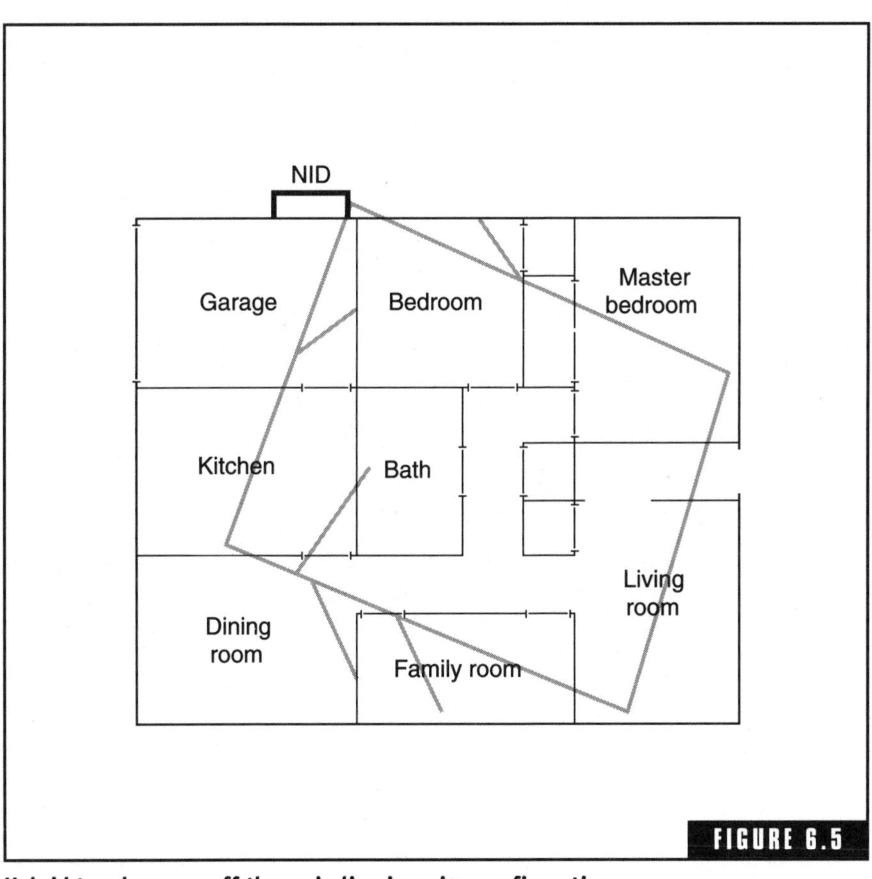

Hybrid topology run off the main line in a ring configuration.

Wiring for Additional Phone Lines

When a new home is built, wiring for multiple phone lines is simple, especially utilizing a star configuration. However, installing additional phone lines to existing homes usually requires some planning and creativity. The phone wiring in most modern homes will support two lines, but often will *not* support three or more lines. Adding a third line will often require adding to existing wiring or completely replacing or reconfiguring the phone wiring throughout the house. If the additional line is to be used as a dedicated line in only one room (e.g., a home office), it is usually a simple matter of installing a separate single line. This type of installation involves running a single cable from the NID to one single-line jack inside the house. Although simple, this type of installation does not take full advantage of having

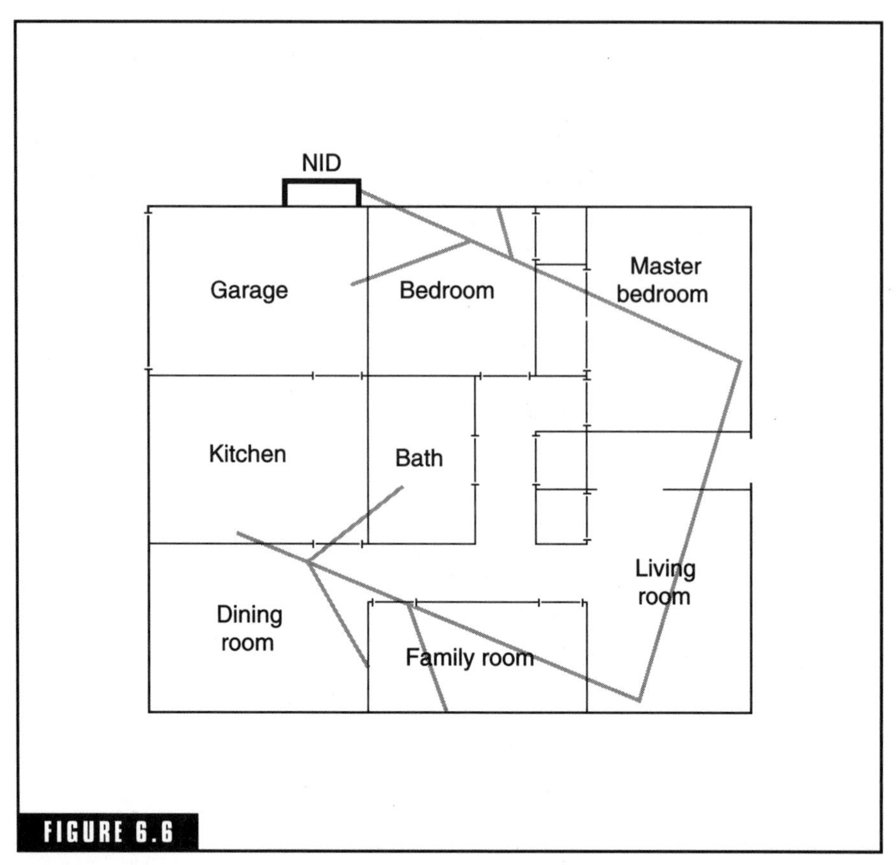

FIGURE 6.6

Hybrid topology run off the main line in a bus configuration.

an additional line in the house. Reconfiguring the wiring into a star topology will provide maximum benefits for all the lines in the house.

Telephone Polarity

The terms used to indicate telephone wire polarities are *tip* and *ring*. These antiquated terms are, in fact, holdovers from the old switchboards where operators plugged in cords to make a connection. The plugs had a tip connection, and just below the tip was a ring. Holdover or not, the labeling system is standard today, with the polarity of each of the pairs identified as follows: Less color is tip, more color is ring. The colors are applied with bands of white. If the band of white is larger than the exposed base color, it is tip. If the white band is small compared to the exposed base color, it is ring.

Two different wire color schemes may be encountered during work on phone wiring. The simplest color scheme is used on normal station cable (what phone technicians refer to as *JK*), which has only two pairs of wire. The first pair has one green wire (tip) and one red wire (ring). The second pair has one black wire (tip) and one yellow wire (ring). For a single phone line, only the red-and-green pair is normally used. The black-and-yellow pair is normally spare and available to install a second phone line.

The other color scheme is somewhat more complicated and is based on a primary color and a secondary color. The tip wire is mostly the secondary color with marks of the primary color (i.e., white with blue marks). The ring wire is mostly the primary color, with marks of the secondary color (i.e., blue with white marks). Primary colors are blue, yellow, green, brown, and slate, or gray. The secondary colors are white, red, black, yellow, and violet, or purple if you prefer. The color code organization for category 3, four-pair communication cable is divided into four colors—blue, orange, green, and brown—to identify each of the four pairs. Table 6.1 indicates the color code identification for category 3 wire.

Connecting Block

The key to a versatile installation is to have a central distribution point that serves as the center of the star configuration. Every telephone circuit will come to this distribution point. It serves to facilitate initial connections as well as reconfiguration. It is usually located in a utility closet or garage wall near the NID (what the phone company calls the point of demarcation). The hardware for the central distribution point is called a *connecting block.* Several types of blocks are available.

TABLE 6.1 The Color Code Identification for Categories 3 and 4

Standard four-pair UTP wire color codes		
Pair 1	Tip	White/blue
	Ring	Blue/white
Pair 2	Tip	White/orange
	Ring	Orange/white
Pair 3	Tip	White/green
	Ring	Green/white
Pair 4	Tip	White/brown
	Ring	Brown/white

Surface-Mounted Screw Terminals

These are probably the most common junction blocks found in residential installations. Wires are connected on the screw terminal according to color. Just connect the wires for your new cable to the matching screw terminals, and route the cable to the location for the new jack.

Screw-terminal-style junction blocks are suitable for splices and taps in topologies other than star configurations. However, using screw-terminal-type junction blocks as a hub in star topology usually results in a tangled mass of lines with many potential failure points.

The top-of-the-line junction boxes rely on *insulation displacement connectors* (*IDCs*). These blocks are much faster to use and provide a more reliable connection than binding posts with screw terminals. The IDC systems provide a gastight seal, which prevents bimetal corrosion. Jumper wires are used for cross-connections at the hub, from the cable coming from the NID to the cables going to the jacks. Jumper wires can be rearranged to change the configuration of the lines. This type of reconfiguring without actually changing the cable connections greatly reduces the potential for problems caused by broken wires in the cables.

IDCs require a special punch-down tool, commonly known as an *impact tool and blade.* Do not use common tools, such as needle-nose pliers and screwdrivers, for punching down connections. Connectors can be damaged so that they appear to work, but are in fact unreliable. The gastight connection is compromised, which results in ongoing reliability problems. A successful IDC requires the right amount of force applied to the right place.

Type 66 Blocks

The most commonly used connection blocks employ the industry standard type 66 IDC punch-down clips. A standard configuration is referred to as the *M block,* an example of which can be seen in Fig. 6.7. The M block is equipped with 50 rows of four slots and will accommodate 25 pair of wires. If more capacity is needed, more blocks can be added. These M blocks serve as a patch panel in which incoming phone company lines come in one side and the wires going to each room in the house attach to the other side. The telephone company lines can be connected to any of the lines going to rooms. With four slots on each row, there is a lot of flexibility for hooking up different configurations. You can also cross-connect computer networks on M blocks.

Cables are normally terminated on the outside row of punch lugs while cross-connects are made on the inner lugs. This avoids accidentally cutting the cable wires when placing cross-connects. The tip is terminated on the top lug of a pair, and the ring on the lower lug of a

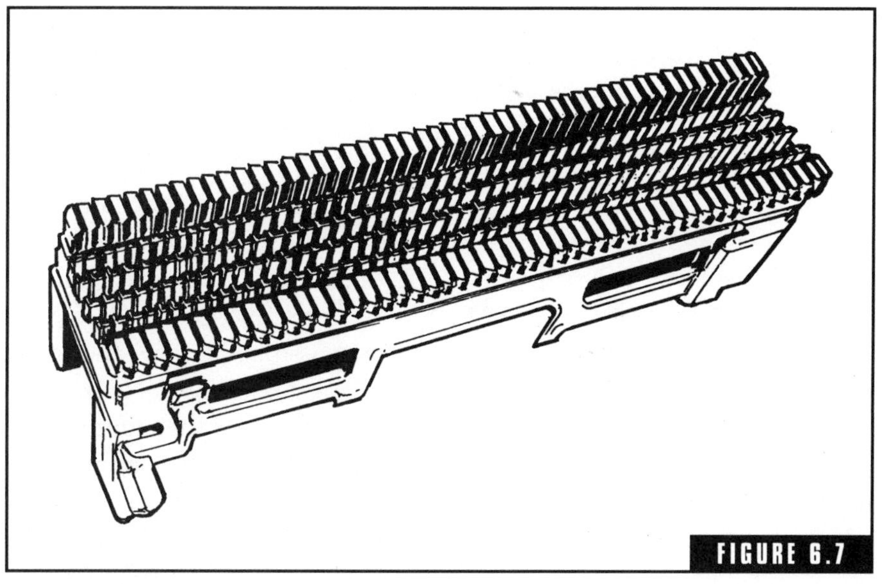

FIGURE 6.7

Example of an M block.

pair. All pairs of a cable should be terminated, even if they will not be used immediately. This allows the addition of lines at that location in the future or allows pairs to be switched if one is broken somewhere in the cable. If a mix of four-pair and two-pair cable is used, reserving the two positions after each two-pair cable permits future expansion of that jack location to four-pair by either replacing the cable or running a second two-pair cable.

Running the Lines

The first concern is to get the standard phone cable from the NID to the 66 block. Normally, one cable carrying at least four twisted pairs of wires is recommended. If two-pair cable is used for this connection, make sure to label each cable as to which is used for lines 1 and 2 and which cable will be used for line 3.

If the house was originally wired with two-pair cable, some jack locations will require more pairs than are available in the existing cable. Always figure on running new four-pair cable to all locations. Where access to three lines is required, install two-jack faceplates with four cable pairs. Pairs 1 and 2 of the cable are connected to the top jack, and pairs 3 and 4 are connected to the bottom jack. A special jack faceplate, which would allow each pair to be placed on a separate jack, may be used for some locations. This type of jack is recommended when each line is used for a different purpose, such as line 1 for voice communications, line 2 dedicated for fax use, and line 3 dedicated to a modem.

Ensure the correct connection of the tip and ring. Pairs 3 and 4 are the most confusing. Remember that tip white/green connects to tip green, ring green/white connects to ring red, tip white/brown connects to tip black, and ring brown/white connects to ring yellow. When the wiring is completed as described above, the home will actually be wired for four lines. Simply placing the desired cross-connects on the 66 block of the star hub will connect the fourth line. For homes that require four or more lines, consider installing a key system to manage access to the various lines and maximize flexibility. It should be a relatively painless process to connect a key system in a home wired in a star topology with four-pair cable designed for three or four lines.

Dual Coaxial Cable Wiring Systems

Probably the best option for future-proofing residential installations uses a cable system known as dual coaxial cable communications wiring. This system is comprised of a wall plate that includes two coaxial connections and an RJ-45 jack with four twisted pairs of low-voltage wiring. One coaxial cable is designated to carry signals from the head end to output devices such as the televisions, VCRs, and computers. The other coaxial cable is designated to carry signals from devices such as CCTV cameras or VCR, laserdisc, or computer back to the head end.

Several manufacturers offer dual coaxial cable systems. U.S. Tec, IES Technologies, and Greyfox Systems offer complete systems designed for compliance with CEBus standards. Molex and AMP offer dual coaxial cable systems under the Smart House brand name. Although Smart House is not currently designed to comply with the CEBus standard, it should work for most needs; however, installers are required to take Smart House training and certification classes to buy or install its products at this time. Keep in mind that Molex and AMP wall plates are only designed for use with other Smart House receptacles and components, while the U.S. Tec, IES, and Greyfox wall plates work with standard and Decora-style duplex receptacles.

A benefit with any of these systems is that they come bundled with telephone wiring, which connects to the same wall plates along with the coaxial cable connections. With this arrangement, users can plug a telephone jack, coaxial connection, or both into any wall plate. Since many locations require the types of wiring these systems provide, they are a good option for communications wiring.

The telephone outlet on these wall plates is an RJ-45 jack with eight wires. The center two pairs of conductors are used for two residential telephone lines. In compliance with the CEBus standard, U.S. Tec, IES, and Greyfox use pins 1 and 2 for an infrared repeater system. This offers a simple way to conduct infrared remote control signals to equipment in other rooms. An infrared receiver plugged into a telephone jack in one room conducts the signals over its pair of low-voltage wires to other rooms where an infrared emitter reproduces the same signal. The Molex and AMP systems propose that all eight pairs of low-voltage wires in their jacks are for use as four residential telephone lines, but a pair of their wires can also be used in this manner.

Networks

The term *network* applies to connecting computers, associated peripheral equipment, and/or communications devices to allow file, program, and equipment sharing. Until recently, networks have been installed for commercial and industrial uses. However, with the increase in home-operated businesses, networks are finding acceptance in residential construction as well.

More than one person is often involved in running a home-based business. More than one person working in the home may require more than one computer; certainly this would require more than one phone, and often an intercom of one type or another would be useful.

A computer network or data communications network confined to a room, building, or group of adjacent buildings comprises a local-area network (LAN). Using this type of network in connecting devices at greater distances (say, from home to the office downtown) changes the name of the network from LAN to WAN (wide-area network). LANs contain specific networking equipment designed to facilitate communications over short distances. Because of the short distances involved in a LAN, this networking equipment operates at high transmission rates for relatively low costs.

One of the primary benefits of LAN equipment is that normally LAN provides faster data transfer than even the fastest phone line modem (roughly a million times faster). A LAN ties two or more personal computers together, allowing them all to use a single printer, fax machine, or copier. Another use of LANs is to designate one of the personal computers as a file server. *File server* is the term given to a designated computer in the LAN. The file server is set up so that other computers can access its hard drive as if it were their own. This permits the computers in the LAN to share files and programs. For example, the file server would have the primary programs (such as word processing or accounting packages) on its hard drive. Other computers in the LAN would download the needed program from the server. Two or more people working on different segments of the same project can share files with one another without copying files to a floppy disk and physically taking it to another workstation. Workers can also send and receive e-mail between the computers in a LAN.

Wiring for Networks

Connecting equipment to a LAN and tweaking all the interfaces and software to the point where the network performs as desired by the homeowner are beyond the scope of this book and, in most cases, beyond the readers' responsibilities. The good news is that the prewiring already covered in this and previous chapters can easily be used for a local-area network. A relatively new ANSI wiring standard, TP-PMD, allows for 100 megabits per second (Mbits/s) of data transmission over unshielded twisted-pair (UTP) copper wiring. TP-PMD makes it possible to establish networks over category 5 UTP cable, as opposed to using fiber optics, which cost about one-third more than category 5.

As residential networks become increasingly popular, many contractors will be more involved with LAN architects and installers during the planning stages. In the following section, a brief overview of some LAN terms and systems provides some useful information for these planning sessions.

Ethernet

One of the most popular LAN configurations used today, Ethernet breaks data into packets which are transmitted via an algorithm until they arrive at the specified destination. The bandwidth (the amount of data that can be handled each second) is about 10 Mbits/s.

Ethernet cables are classified in an *XbaseY* format, where *X* is the data rate in megabits per second, *base* refers to baseband (as opposed to radio frequency), and *Y* is the category of cabling. Several types of cables are common:

- *10base5:* This is the original "full-specification" variant of Ethernet cable, which uses a stiff, large-diameter coaxial cable with an impedance of 50 ohms (Ω) and with multiple shielding. The outer sheath is usually yellow so it is often just called *yellow cable*. It is designed to allow transceivers to be added while existing connections are live by using a *vampire tap*. A vampire tap clamps onto the cable, forcing a spike through the outer shielding to contact the inner conductor while other spikes secure the tap to the outer conductor.
- *10base2 ("thinnet")*: This variant of Ethernet uses thin coaxial cable (RG-58 or similar), as opposed to 10base5 cable (see Fig. 6.8).
- *10baseT:* This is a variant of Ethernet which allows stations to be attached via twisted-pair cable, which is now very common.
- *100baseT:* Increasingly common, this is the predominant form of Fast Ethernet. 100BaseTX runs over two pairs of wires in category 5 cable.

Homes prewired or rewired with category 5 twisted-pair cable can handle the fastest current network technology and will keep up with advances in the future.

Fiber Optics

We have been hearing about fiber optics for quite some time now, and some areas of the country are now experiencing the first real implementation of this technology. A plastic or glass (silicon dioxide) fiber no thicker than a human hair is used to transmit information by using infrared or even

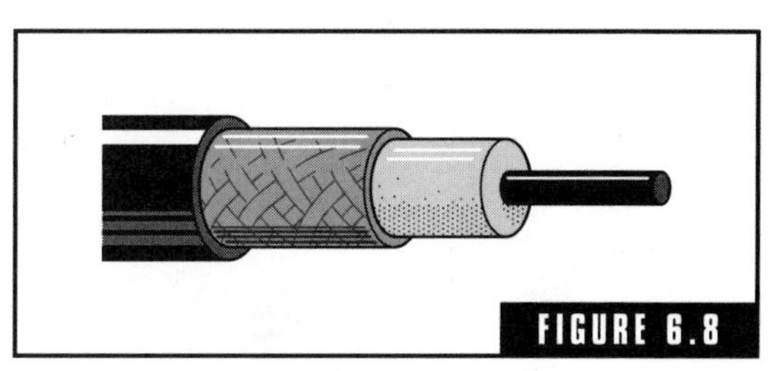

FIGURE 6.8

10Base2, "thinnet."

visible light as the carrier. A laser usually generates the carrier light. The information transmitted includes audio, visual, and data. Audio transmissions can be voice, music, or computer-generated tones. Visual transmissions range from graphics and photographs to television programming and even live video signals from one personal computer to another. Live video capabilities make fiber optics ideal for video conferencing in which two or more people at separate sites can have a face-to-face meeting on their computers.

Optical fiber (Fig. 6.9) is less susceptible to external noise than other transmission media, and is cheaper to make than copper wire, but it is much more difficult to connect. Optical fibers are difficult to tamper with, making them appropriate for secure communications. The light beams do not escape from the medium because the material used provides total internal reflection. The potential of fiber optics is staggering. AT&T Bell Laboratories has sent information at a rate of 420 Mbits/s through an optical fiber cable. At this rate, the entire text of the Encyclopaedia Britannica could be transmitted in 1 s. A single fiber can transmit 200 million telephone conversations simultaneously.

Serial Interface

Data communication interfaces, which carry digital information, can be classified into two broad categories: serial and parallel. Serial sim-

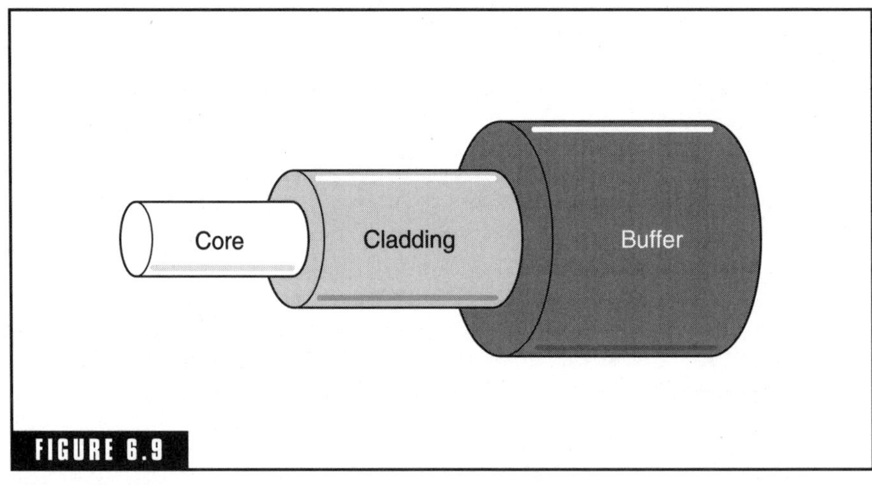

FIGURE 6.9

Fiber-optic cable.

ply means that the data are sent a single bit at a time, whereas parallel means that groups of bits are transferred simultaneously over multiple paths operating in parallel. The tradeoffs are lower-cost wiring for serial versus potentially higher data transfer rates for parallel. Serial interfaces are more applicable for long-distance communications due to lower media costs, and these interfaces avoid the timing-difference problems common to multiple parallel simultaneous signaling paths.

This section focuses on several serial interfaces. RS-232 and RS-485 have been around for awhile and are quite well known. A third, IEEE 1394 (a.k.a. Firewire, a trademark of Apple Computer), is relatively new but has been receiving considerable attention.

EIA RS-232 Interface

The Electronic Industries Association (EIA) developed the RS-232 interface standard in the late 1960s. The standard has seen several updates over the years, and the present version is officially known as "ANSI/EIA/TIA-232-F, Interface between Data Terminal Equipment and Data Circuit-Terminating Equipment Employing Serial Binary Data Interchange." There is also an international equivalent: "ITU Recommendation V.28, Electrical Characteristics for Unbalanced Double-Current Interchange Circuits."

RS-232 has quite a long definition, but do not be intimidated by it, for we will explain it all. RS-232 is a point-to-point single-ended signaling binary baseband serial interface that supports either synchronous or asynchronous communications, in simplex, half-duplex, or full-duplex modes. Point-to-point means that two, and only two, devices are connected. Single-ended signaling means that the electric signals are carried over a single wire with a separate ground return path. Binary and baseband mean the data to be communicated have only two defined values (1 and 0) which are signaled as two distinct voltages on the wire. Synchronous means that the timing references at the transmitting and receiving ends of the connection are kept in step by additional clock synchronization signals. Asynchronous means that the timing references at each end of the systems are not coordinated. The timing references must rely on the receiving device's being able to accurately determine the beginning of a block of data, along with having an agreement on the signaling speed. Simplex means that the data always go in one direction only. Half-duplex means that the data can go in either direction, but not simultaneously. Full-duplex means that data can be sent in both directions simultaneously. As seen in Fig. 6.10, the RS-232 industry standard for data transmission is a compatible, reliable method to interconnect external functions to a computer. RS-232 defines two distinct ends to the interface: the DCE and the DTE. These are the data circuit-terminating equipment and data terminal equipment in the title of the standard. Examples of DCE-DTE pairs are a modem and computer or terminal, or a modem and printer.

The RS-232 electric signals swing around ground. The minimum, fully loaded, driver output voltage is ±5 V, while the maximum, unloaded, is ±15 V. The drivers must be able to withstand short circuits to any other line in the interface. The receivers are required to have thresholds within a ±3-V window. Both the drivers and receivers are required to be able to withstand ±25 V. These signaling voltage levels provide a minimum noise margin of 2 V. The maximum signaling

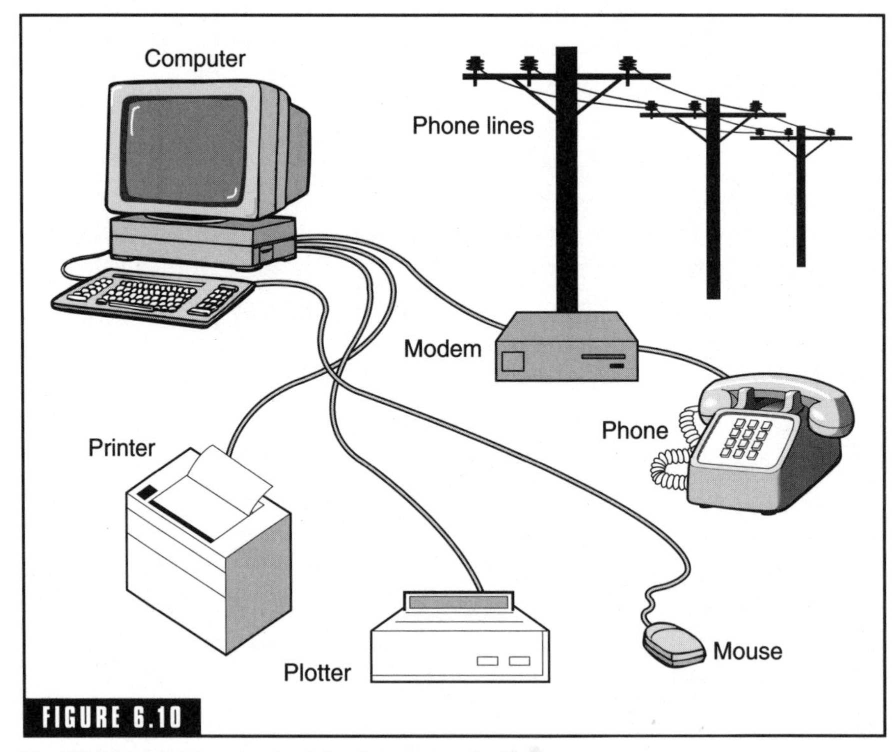

FIGURE 6.10

The RS-232 industry standard for data transmission.

rate covered by RS-232 is 20 kilobits/second (kbits/s); other compatible standards cover higher rates.

RS-232 defines a 25-pin connector, commonly referred to as DB-25, and a less common but smaller 26-pin connector. Some additional connectors, notably an 8-pin and 9-pin, have been defined by other related standards. Although most of the pins have specific uses, many of these are optional. In addition, the DTE connector is male while the DCE connector is female. The maximum length of the cable is not specified, but rather the maximum total load capacitance is limited to 2500 picofarads (pF). This translates to about 50 ft for inexpensive cables.

Data format, data rates, and signaling protocols are not defined in RS-232. These parameters may be defined by the other standards such as for the DCEs or modems. The implementation of these functions is handled by an additional logic circuit, called a UART or USART. These terms stand for universal asynchronous receiver/transmitter and universal synchronous-asynchronous receiver/transmitter. Either switches or software is commonly used to set the required communication parameters. Either way, if there is a parameter mismatch between the two ends of the interface, it is unlikely that the system will work.

Since the driver output voltages specified in RS-232 swing both positive and negative relative to ground, it can present a bit of a problem for many electronic systems, which may be operating with only a positive polarity supply. Even more problematic is the trend toward lower supply voltages, such as 3.3 V, which makes generating the required ±3-V minimum driver output a challenge. Fortunately, integrated circuits (ICs) are now available which include charge-pump circuits that can provide both of the needed positive and negative operating voltages from a single 5- or 3.3-V source.

The dual-polarity driver output voltages also present a challenge for circuit integration. Although the speeds and voltages are easy for discrete component implementations, trying to integrate them with single supply logic circuits requires special processing which results in less than optimal circuit densities and higher-than-needed costs. In general, only a limited amount of logic is included on the RS-232 integrated circuits. The good news is that some of the RS-232 ICs have now been in production for more than 25 years and are very inexpensive.

One of the more obvious limitations of RS-232 is the signaling rate. The standard itself only covers rates to 20 kbits/s, although higher rates are possible and common on newer equipment. A major reason for this limitation is the potential for crosstalk between the signal lines in the cable. Slowing the rise and fall times of the signals controls this interference, but becomes a limitation on the speeds for the bits being signaled. A recent standard, TIA/EIA-694, has been introduced which maintains backward compatibility with RS-232 but raises the signaling rate to as high as 512 kbits/s. Despite RS-232's age and some of its less-than-ideal characteristics, it is still in wide use. With all the options described above, it is easy to see why there can be confusion or problems in setting up RS-232 communication links. Today the most common use of RS-232 is for asynchronous, telephone modem type of communication on personal computers. Virtually every PC has at least one RS-232 "serial port," and most desktop systems have two. The RS-232 used on PCs only uses 5 of the original 23 signal lines. A nine-pin connector definition has also been added. In this use of RS-232 only short 7- or 8-bit-long blocks of data can be reliably transferred before resynchronization is needed.

RS-485 Interface

The need for faster communications rates and longer distance capabilities led to the development of numerous standards beyond RS-232. Most of these newer systems use an approach to electric signaling called *balanced differential mode.* Balanced differential mode implies the use of a matched pair of wires to carry complementary, or mirror-image, signals. The advantage of this system is primarily the noise immunity. The noise that will be picked up will be essentially the same on both signal wires, so a well-designed differential receiver circuit can separate out the difference (desired) signals from the common-mode (noise) interference. Manufacturing fast differential circuits is fairly easy with ICs, but is very difficult with discrete components. The common-mode (noise) range of RS-485 is from −7 V to +12 V, which is much better than that for RS-232.

The present, newly revised version of RS-485 is "ANSI/EIA/TIA-485-A, Standard for Electrical Characteristics of Generators and Receivers for Use in Balanced Digital Multipoint Systems." RS-485 supports transmission rates to 50 Mbits/s, distances to 1200 meters (m), a multipoint bus architecture with 32 nominal devices, in a half-duplex communications mode. Multipoint bus means that the transmission line wires provide direct electrical connections to more than just two devices. The actual number of devices attached to the bus is dependent on their electrical impedance characteristics. Also, there are tradeoffs between the distance and the data rate, so that achieving 50 Mbits/s at 1200 m just is not possible.

IEEE 1394

IEEE 1394, also known as *FireWire,* is the newest and fastest serial interface, with the ability to transmit data up to 5000 times faster than RS-232. The work on this standard actually started as a definition for a relatively low-speed auxiliary control and communications channel for some high-performance internal computer "backplane" multipoint busses. A desire to link multiple systems together using a cable, along with the realization that the data rate could be increased, led to a shift in focus in the early 1990s. As the interest in cable links grew, it became clear that a system using small connectors and cables, capable of high data rates, would make a good communications bus for connecting personal computers, peripherals, and multimedia audiovisual equipment. A particularly strong motivation was the shrinking size of portable PCs, where the overabundance of connectors was beginning to be a serious limitation on how small the case could get, or how it could be shaped. What evolved out of this effort became an official standard of the Institute of Electrical and Electronics Engineers (IEEE) in December 1995, IEEE Standard 1394-1995 for a high-performance serial bus.

The FireWire interface standard is considerably more complex than either RS-232 or RS-485. The documentation is more than 380 pages, compared with about 45 pages for RS-232 and about 26 pages for RS-485. In order to manage this higher level of detail, the FireWire standard is broken down into several layers:

- Cable physical
- Backplane physical links
- Transaction
- Management

Each of these is covered in a separate "clause" or chapter; three clauses cover introductory information, and there are 12 annexes providing additional information.

Of the two physical layers, backplane and cable, only cable will be covered here. The backplane layer primarily applies to computer and communications applications. The cable physical layer is more commonly used in consumer applications, so it will be the focus of the rest of this section.

The most obvious features of FireWire are the connectors and cables. Equipment having FireWire cable connections, or ports, may have between 1 and 27 of these ports, with 1 and 3 being most common. The cables used to connect these ports have two individually shielded twisted-pair transmission lines, and they may carry another pair of wires for auxiliary power and power return. There is also an overall shield. The port connections provide impedance-matched terminations for both ends of each pair, plus drivers and receivers for each pair. The individual cable links are used in a point-to-point configuration. Logic in the transceiver chip, when there is more than one port, performs a "one-in-to-many-out" signal-repeating function to provide the shared multipoint bus behavior.

The topology in which FireWire devices may be cable-connected has only a few restrictions:

- The maximum number of devices on the bus is 63.
- The maximum number of cable "hops" allowed between the most widely separated devices is limited to 16.
- No loops are allowed.

Within these limits, devices may be daisy-chained together or connected in a branching tree or star pattern.

Most of the remaining interesting features of FireWire are not physically obvious. They are the results of implementing the protocols specified in the layers of the standard above the physical layers. Some of these features include automatic bus configuration, fully defined packet formats, and complete control over the flow of data. The details of how these higher-level protocols work are beyond the scope of this book. The benefits to the users are hassle-free hookup. There is no in/out or up/down direction sense in FireWire. It makes no difference which socket is used on a particular device, and the order in which devices are connected together is irrelevant.

There are two distinct operating modes in FireWire: During initialization and arbitration the bus operates in a nonclocked full-duplex mode, while during data transmission it operates in a clocked half-duplex mode. The data transmission modes are defined at three different signaling rates; these are termed S100, S200, and S400 and run at 98.304, 196.608, and 393.216 Mbits/s. The data transmission phase supports two types of packet delivery: asynchronous and isochronous. The first is a high-reliability mode with acknowledgments, or retransmission if there was an error. The isochronous operation provides guaranteed access for delivery of time-critical data such as video or audio streams.

The main consumer use for FireWire is as the digital interconnects for the DV-format camcorders, VCRs, and video editing systems. These need to be able to handle sustained uniform data rates of 30 Mbits/s. The use of FireWire for this digital interface allows completely loss-free editing and copying of tapes, which is impossible with an analog interface.

The future for FireWire is linked to the proliferation of new services such as digital satellite and digital cable and new products such as digital televisions and digital versatile discs (DVDs). A number of standards bodies are actively defining how these digital data will be handled over this protocol.

Of greater interest are the prospects for FireWire in the home. Low-speed home automation control signals are now handled quite well by power line carrier or dedicated RS-232 and RS-485 systems. Rather than use FireWire as the network for carrying such low-speed signals, bridge devices will enable communications between the different systems. This will allow home automation commands to be originated in digital TVs or PCs, and for status messages to be returned to those devices.

Currently, using FireWire to connect digital audiovisual and PC equipment together between rooms might seem to be a problem because of the 4.5-m cable hop length. Several groups are now actively working on new physical layers that will be able to provide single hops of 75 to 100 m. One of these efforts will allow the use of readily available and inexpensive category 5 UTP wire, already a common part of structured wiring systems. This will be at FireWire's 100

Mbits/s base rate, and is targeted at home networks. So category 5 wiring can be installed today and provide for the digital home network of the future. Higher data rates for FireWire, initially at 800 Mbits/s and 1.6 gigabits per second (Gbits/s) are also being defined. Glass fiber will most likely be used at the higher rates, but the need for that much capacity in a home seems to be quite far in the future.

Overlapping Technologies

Currently, the most used buzzword in high-technology circles is *convergence.* Simply put, convergence refers to combining two or more traditionally autonomous technologies into a new hybrid. Most convergence discussions center on melding the Internet with television to create interactive TV (discussed in Chaps. 7 and 10), providing new and exciting options and conveniences for the homeowner. However, other products are being developed that have nothing to do with television but will provide convenience, cost savings, and practical applications for the home's occupants.

Phone company rate plans, aside from being numerous, are frequently complex and convoluted. The number of small companies offering even lower (but widely varying) rates increases the confusion, making it very difficult to decide which carrier to use for any particular call. Many people surrender the choice and use the default carrier, which was automatically assigned along with the local service. This situation presents a perfect opportunity to utilize the unique abilities of a personal computer.

MediaCom has produced a product called PhoneMiser, which puts the PC's power to work in making the choice of telephone carrier a little easier. PhoneMiser contains both hardware and software. The hardware is a device that is put in-line between the parallel port and the printer of a Windows 95–based PC, and is also connected to the phone line used for voice calls.

PhoneMiser's software includes a lookup table of rate structures for a number of large and small telephone companies. MediaCom updates this information each month through a direct connection with the computer's modem. The device constantly listens in on the telephone line. When a toll call is dialed, the program notes the number and interrupts the call before it goes through. The number called is compared against the table of rate structures to determine the least expensive routing for that particular call (domestic or international). Special rate plans, time of day, day of the week, and any other factors that determine call rates are included in the process. From all the available carriers the program selects the least expensive carrier, and within 1 second of the call's being dialed, reroutes the call to go through the chosen carrier.

This application is both a prime example of convergence and a defining use of home automation. Remember that the rigid interpretation of the term *automation* means that no conscious thought or efforts are needed for the desired action to be carried out. This use of a specialized computer program meets those standards of automation, no thought or additional effort is needed for the desired result (selecting the least expensive carrier) to be accomplished.

Currently, some areas of the United States are experiencing the deregulation of another utility, electric power suppliers. As happened with the telephone industry, electric companies are now competing for consumer business. Various rate plans are being sent to homeowners along with reams of public relations information. Once again, this influx of choices and confusing legalese serves to confuse the average homeowner. Certainly the future will yield a system to automatically select the least expensive electric provider at any given time and day.

Yet another example of converging technologies is being offered by many telephone companies as an optional extra. A system known as *BusCall* uses global positioning system (GPS) technology to keep track of individual school buses and to notify each subscribing home on the route 5 minutes prior to the bus's arrival. The buses carry an on-board GPS satellite receiver, and its location is automatically sent to the BusCall processing center via data-over-cellular transmission. About 5 minutes before the bus will actually reach a subscriber's school bus stop, the system calls the subscriber's phone with a voice message. This message announces the arrival of the bus, including the number of the bus and the estimated time of arrival. This notification can alternatively be sent via a pager or e-mail.

It might be interesting if this service could be expanded to provide positioning information on fleets of construction vehicles.

Larger contractors could know where each truck, bulldozer, backhoe, and grader are at any given time of day or night. Not only could this service make scheduling and supervisory chores easier, but also in the event of a theft, it could prove invaluable to recovering stolen property.

More information on convergence technologies can be found in Chaps. 1, 7, and 10.

Notes

Notes

CHAPTER 7

Entertainment

Manufacturers seem to be introducing new entertainment components daily. Many homes already have multiple televisions, one or more videotape recorders, receivers, cassette decks, compact disc players, telephones, and some kind of answering machine or voice mail. New technology is being introduced which makes some of these systems seem insufficient while enhancing the performance of others.

As the demand for new technologies increases, contractors and builders will need to be familiar enough with these systems to discuss the layout of media rooms with home buyers. Installers, while being part of the consulting process, will also need a deeper knowledge of system components in order to advise homeowners on the type(s) of components, systems, and distribution of home entertainment. The information in this chapter introduces some of the major components, and there is discussion of system design and integration with home automation.

Home Theater

With the introduction of premium services (HBO, Showtime, etc.) on cable television, people discovered the pleasures of watching full-length, uninterrupted movies in the comfort of their own homes. Affordable videotape recorders added another benefit to home viewing by putting the power of program scheduling into the hands of the homeowner. Anyone with a VCR could now go to a rental store and select the evening's (or weekend's) entertainment from the available movies. To further enhance the viewing experience, VCRs were then equipped with audio/video (A/V) jacks that connect the VCR to a stereo receiver. This connection pipes the soundtrack through the homeowner's stereo system, providing richer, fuller, and louder sound. Since those early days, manufacturers have been trying to reproduce the theater experience in the home.

Recent surveys have shown that the number one item on a homeowner's wish list has become a home theater system. This desire applies to new construction as well as existing homes. Movie theaters are intended to involve the viewer in the action as the filmmaker intended. Home theater strives to be true to the filmmaker's efforts as well, and a properly designed system with good components can actually surpass the movie house experience.

The first decision in the planning process focuses on which room to designate as the home theater. A carefully installed entertainment system is unobtrusive so that the room doubles for other uses. The family's lifestyle and viewing preferences will play a determining factor in the choice of room. Just about any room can be used as a home theater, but there are some considerations.

Acoustics

Audio experts have known for years that the physical attributes of listening rooms are every bit as important as the audio equipment involved. As a matter of fact, some would argue that room acoustics are far more important than popular equipment features such as gold connectors, polypropylene capacitors, and the like. Acoustical sound fields that occur in an enclosed space (Fig. 7.1) vary and are quite complex. However, three components outweigh the others:

- *Direct sound:* The direct sound from the speakers themselves is the sound waves that travel in a straight line directly from the speakers to the listener's ears. These direct sounds are probably the most significant component of sound reproduction due to their relatively large amplitude and direct transmission characteristics. Successfully designing a home theater includes supplying an unencumbered line-of-sight path from the speakers to the listeners.

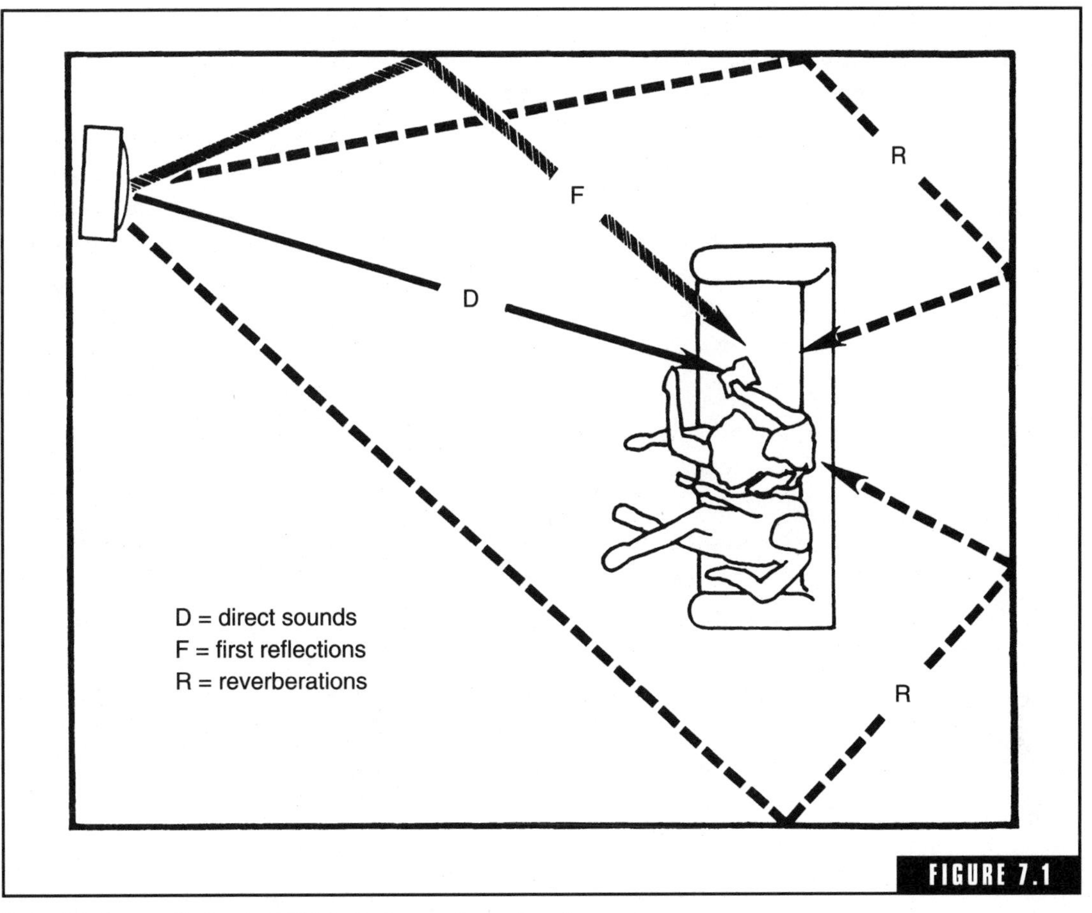

Example of acoustical sound fields that occur in enclosed spaces.

- *First reflections:* First reflections are the sound waves that bounce off the surfaces flanking the speakers and the listeners. These early reflections are important to the perception of the "sound stage," and a failure to properly attenuate them can result in a marked reduction of the *breadth* of the recording.
- *Reverberation:* Reverberation consists of secondary and random reflections that bounce off other surfaces in the room and eventually arrive at the listener's ears. These sound reflections reinforce the feeling of room size and ambience. A room with hard surfaces, such as a bathroom, provides reverberation echoes bouncing against the walls and creating the feeling of a larger space.

These three sound components bouncing around the room combine to form the acoustical signature of the room. This acoustical signature effectively changes the primary audio recording. The listener hears the recording almost as if there were two separate sound systems playing in the room at the same time.

In designing a listening environment it is important that these two "systems" be sonically balanced as well as possible. For example, too many hard surfaces will result in an overemphasis on reverberation. On the other hand, a lack of reflections deadens the room, making the recordings sound lifeless. The ideal home theater combines direct sound with enough first reflections and reverberation to balance the resultant sound, making it sound natural. To balance first reflections and room reverberations, absorption and diffusion principles are used. Absorptive surfaces consist of materials that dampen sound energy so that only a fraction of the energy is reflected. The portion not reflected converts to thermal energy, which dissipates into the air.

Absorption coefficients with value scale ranges from 1 to 0 indicate how well a material absorbs sound. An absorption coefficient of 1 means sound energy is absorbed completely. An absorption coefficient of 0 means sound energy is reflected entirely. Since no existing materials fall at either extreme of the scale, all absorption coefficients lie somewhere between the two extremes. Table 7.1 shows absorption coefficients for typical surfaces found in residential structures.

Diffusion takes the sound waves directed at the diffusive surface and breaks them into many small components. This principle scatters the sound field around the room at a greatly reduced magnitude. Specially engineered diffusion panels, bookcases, and furniture can augment diffusion within the media room.

Floors and Ceilings

Floors and ceilings constructed of extremely reflective materials normally present problems for media rooms. Carpeting is a standard household building material that does a wonderful job of increasing a surface's absorption coefficient. Wall-to-wall carpeting and area rugs do much to absorb sound energy. In retrofit applications, acoustic ceiling tiles actually do a good job of absorbing incident sound waves, particularly in the higher frequencies. Drop ceilings do an even better job of sound absorption. The standard "wormhole" pattern tile is common in commercial applications because it provides a good mix of absorptive and diffusing properties.

Note: High listening levels tend to rattle drop ceilings. Several manufacturers of drop ceilings offer a solution to the problem by providing small rubber hangers that cushion and isolate the structure of the drop ceiling.

New construction and additions should include materials that are decor-friendly and specifically designed for ceiling use. Sonex panels made from melamine absorptive foam, for example, are available in a number of different surface textures and colors. If the look of Sonex does not complement the decor, there are plenty of flat-faced products typically constructed of fiberglass or foam materials. Alphasorb panels constructed of fiberglass sheets wrapped with woven decorator fabrics are offered in 65 different colors.

TABLE 7.1 A Brief List of Absorption Coefficients

	Frequency (Hz)					
Material	**125**	**250**	**500**	**1000**	**2000**	**4000**
Concrete block, unpainted	0.36	0.44	0.31	0.29	0.39	0.25
Concrete block, painted	0.10	0.05	0.06	0.07	0.09	0.08
Glass, window	0.35	0.25	0.18	0.12	0.07	0.04
Plaster on lath	0.14	0.10	0.06	0.05	0.04	0.03
Plywood paneling	0.28	0.22	0.17	0.09	0.10	0.11
Carpet on pad	0.08	0.24	0.57	0.69	0.71	0.73
Gypsum board, 0.5 in	0.29	0.10	0.05	0.04	0.07	0.09
Drapery, lightweight	0.03	0.04	0.11	0.17	0.24	0.35

Walls

Residential room walls often cause acoustical problems since they are frequently made from hard substances such as gypsum board or wood. Since the wall surfaces located near the listener are directly involved in first reflections, it is especially important to properly treat them. For centuries, drapes have dampened sound. Furniture and plants also diffuse sound. For example, an upholstered couch or chair absorbs quite a lot of sound. Placed right at the spot on the wall where first reflections occur, an upholstered couch or chair makes a remarkable improvement in sound throughout the room.

Note: When you speak to the homeowners during the planning meeting(s), make sure to address the type and placement of furnishings.

Professional products can be used to attenuate wall reflections as well. Decorator-styled acoustical wall panels designed in many woven-fabric colors make for easy installations and effective absorption. The remaining untreated surfaces in the room contribute to the reverberatory ambience. The reverberation characteristics of the room are simply tested by standing near the listener's position and clapping loudly. The echo can tell you a lot about the reflective sound signature of the room. If the clap produces a distinct echo, more absorption is required.

Audio

Since the late 1920s, when talking pictures made their debut, the motion picture industry has been improving the sound tracks of movies. In 1941, Disney Studios in its production of *Fantasia* introduced the first multichannel sound track. This six-discrete-channel, high-fidelity process, dubbed *Fantasound* by Disney, triggered a rush of competition to match or better its quality. Just as today, better sound quality (and sound effects) drew audiences.

In the 1960s, the motion picture industry took a huge step forward in sound track quality when Ray Dolby developed a process for recording multichannel sound optically onto movie film. The system was called *Dolby Stereo*; even though it had four sound tracks encoded, old stereo movie theaters and the new four-channel surround theaters could use the Dolby system.

The Dolby System

Since its invention, the Dolby system has become a standard in motion picture and audio sound reproduction. Because of this predominance, it is necessary to understand how the process works.

In the commercial Dolby cinema (Dolby Stereo) sound system, the left and right audio tracks are optically read off the filmstrip and fed to a Dolby Stereo cinema processor. This processor decodes the Dolby Stereo signal into four different channels: front left, center, front right, and surround. A subwoofer channel is split off from the information in the front speakers.

Filmmakers use each of the channels in specific ways. Most of the on-screen dialogue from the actors is placed on the center channel. For the sake of understandable dialogue, the center channel is probably the premier channel. The left and right front channels usually handle stereo sounds, such as music and sound effects; but when the action moves to or from one side of the screen, these channels are used to convey that motion to the audience. For example, when a train runs through the scene from one side to the other, the sound effects will move from channel to channel (side to front to side).

In order to put the audience into the action, the system uses surround channels. The surround speakers provide ambience; weather sounds such as rain or wind piped through these channels subtly sets the scene for the viewer. Just as front channels are used for side-to-side movement, surround channels are incorporated into front-to-back movements.

Home Versions of Dolby

Since so much of the entertainment media coming into the home is created using Dolby technology, it stands to reason that home theater equipment should include the decoding technology to reproduce the sound faithfully. Three incarnations of the original Dolby Stereo currently exist specifically for home use.

DOLBY SURROUND

The consumer equivalent of Dolby Stereo is Dolby Surround or just plain Dolby. VHS high fidelity, laserdisc, digital versatile disc (DVD), and television broadcasts all use Dolby Surround technology. Software such as videotapes is imprinted with the logo Dolby Surround. Hardware such as receivers and preamplifiers carries the logo *Dolby Surround Pro-Logic.*

DOLBY DIGITAL

Dolby Digital (what was originally known as Dolby AC-3) and Digital Surround are the home version of the digital sound formats now being used in theaters. Dolby Digital is the most recent version of the popular 5.1-channel home theater sound system. It consists of front left and right speakers, a center speaker, left and right surround speakers and a *low-frequency effects* (*LFE*) channel, usually used with a subwoofer. Laserdisc and digital versatile disc often employ Dolby Digital during recording.

DIGITAL SURROUND (DTS)

DTS is yet another home version of the digital system used in movie theaters. Also known as DTS Coherent Acoustics, the system uses the 5.1-channel sound format for use in home theaters. Left and right front channels, left and right surround channels, a center channel, and a low-frequency effects channel, usually in conjunction with a subwoofer, are used in the system. DTS is available on laserdisc as well as on music-only compact discs (CDs).

Lucasfilm THX

THX is not a sound format; it is a home theater system consisting of audio components carrying the Lucasfilm THX logo. The system is designed to enable the listener to experience film sound as close to the original sound as heard by the filmmakers in the dubbing studio. An audio component with this logo means that it meets the design and performance standards set by THX. There are some design principles of THX whose application can be beneficial to home theater:

- *Front-channel reequalization:* THX certified surround processors have a rolloff in the high frequencies to account for perceived spectral differences in sound between a living room and a large cinema.
- *Surround channel decorrelation:* This is especially useful in Dolby Surround, where the surround channel is monoaural. THX processors divide this monoaural channel into left and right and then introduce *decorrelation* by varying the phase between them. This is meant to make the surround sound more diffused when played back with two speakers.

- *Matching front speakers:* THX requires that the three front channels be identical, to guarantee that sounds panned along the screen will follow a consistent sonic path between the speakers.
- *Dipole surround speakers:* The surrounds in a THX system are not of the conventional direct radiating type. The *dipole* consists of two sets of drivers opposed to each other, radiating sound to the front and rear walls rather than directly to the listener (Fig. 7.2). This is intended to reproduce the diffused surround sound heard in a theater.

THX certification applies to surround sound processors, receivers, amplifiers, and speakers. Bear in mind that there is now THX 5.1, a new set of requirements for six-channel digital sound playback, which only affects surround processors and receivers.

Components

The basics necessary to bring a theatrical experience into the home would include the following components:

- Five speakers—three for the front (left, right, and center) and two for surround (left and right)—are needed.
- A receiver with Dolby Surround Pro-Logic processing and ample amplification for the speakers are needed.
- A high-fidelity VCR will suffice, but a laserdisc player and/or DVD player will add greatly to the system.
- A wide-screen video monitor is needed, as large as the room and the budget will allow.

A more faithful and exciting rendering of the theater experience naturally requires a more extensive system. Recommended components include the following:

A TONALLY MATCHED, THX CERTIFIED FIVE-SPEAKER SYSTEM

Tonal matching between the speakers is particularly important for the front channels, in order to maintain a consistent sonic image for the picture. Seven-channel systems are also available. The principle

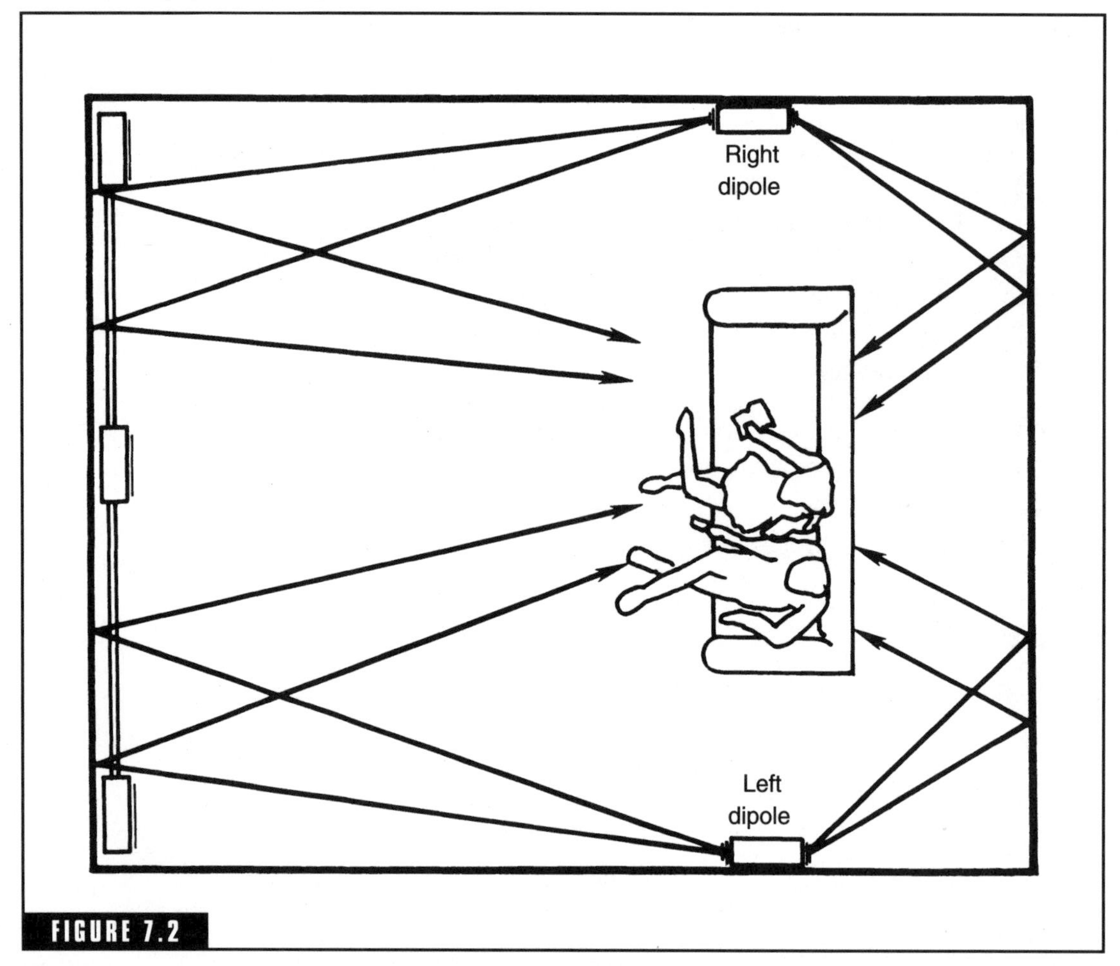

FIGURE 7.2

Dipole systems use two sets of drivers opposed to each other to radiate sound to the front and rear walls rather than directly to the listener.

involves creating left and right surround channels by extracting stereo information from the front left and right channels. The processor interpolates these signals between rear and side speakers. This arrangement can simulate many discrete digital film sounds, including the front-to-surround channel pan along one side of the room.

Loudspeaker manufacturers are now making tonally matched, center-channel versions of full-size speakers for use on top of a video monitor. Stay within the same brand and same family of speakers to maintain consistent timbre in home theater surround sound.

A SUBWOOFER

Subwoofers reproduce the rumbles and deep-bass signals of the soundtrack and special effects.

TACTILE TRANSDUCERS

Tactile transducers are electromechanical drivers designed to couple low-frequency audio energy to physical surfaces, such as couches, chairs, or even floors. They allow the feel of special effects to pass from the system to the viewer. The earth-shaking effects of dinosaurs walking through a scene will result in the transducers shaking the viewer. The audience will feel the rumble and roar of a fighter plane leaving the flight deck or terrorists and monsters exploding skyscrapers.

AMPLIFICATION

A separate preamplifier with surround sound processing and power amplifiers provides better performance than just receivers and allows easier upgrades. The dynamic range and volume needs for cinema sound require receivers and amplifiers which can yield at least 80 watts (W) for the front channels. Normally, similar power should be available to the surround channels. For the subwoofer, wattage requirements are crucial because a lot of power is needed to generate the large cone excursions for low bass output. Many subwoofers have a built-in amplifier. Subwoofers should have a minimum of 100-W amplification. If the subwoofer is not powered, a power amplifier delivering at least 100 W to it is mandatory. More power is generally better, so don't skimp.

DOLBY DIGITAL AND/OR DTS DIGITAL SURROUND

This is the home theater equivalent of the 5.1-channel digital sound systems used in movie theaters.

PROJECTION MONITORS

Projection monitors made specifically for home theater uses allow larger viewing areas than even the biggest television sets. Larger viewing areas permit more of each frame of the movie to fit onto the screen. Movies shown on television are compressed to fit the smaller format of television. More information on this topic can be found later in this chapter.

Speaker Arrangement

A typical speaker arrangement that works well in most home theaters is shown in Fig. 7.3. Place the front speakers in the same plane or hemisphere as the video screen, to achieve direct sound. Place center speakers on top of the video monitor, flanked by the left and right speakers. Large screens for video projection permit all front speakers to be placed below or behind the acoustically transparent screen. If the center speaker is above or below the screen, use shims to angle the speaker toward the listeners.

In an ideal situation, surround speakers are mounted 3 to 6 ft above and 5 ft to the side of the listening position. The principle is to position the speakers relative to the primary listening position so that the surround sound is diffused, not localized. The idea is to not be able to pinpoint their location by listening. If the sofa is on the back wall, mount the surround speakers on the sidewalls in either corner. Make sure to face them toward the listening position. To prevent reflections, move the speakers in front of the TV. Keep 2 to 3 ft of space around each enclosure.

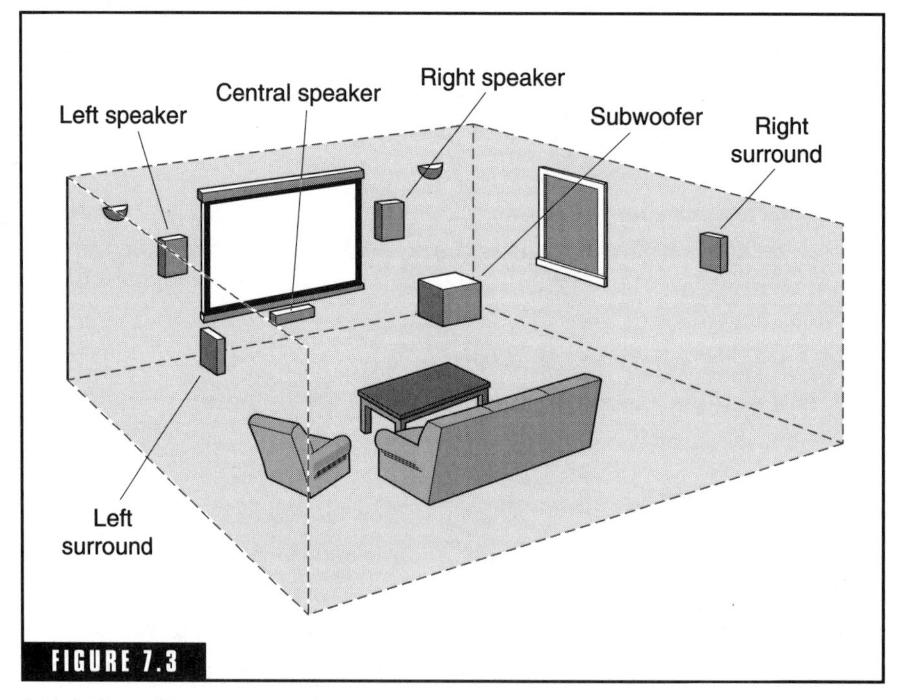

FIGURE 7.3

A typical speaker arrangement.

For subwoofers, several considerations come into play. To cleanly reproduce a given frequency, the largest dimension of the room must equal one-half of the wavelength generated by that frequency. Unless the room is large enough to support very low frequencies [as low as 20 hertz (Hz)], standing waves will result. Standing waves, which result from two parallel boundaries in a room, are actually audible at 200 Hz. Subwoofers come tuned for various frequency responses with 20 Hz being the lowest.

Tip: Use the formula

$$\text{Wavelength} = \text{speed of sound (1100 ft/s)/frequency}$$

to calculate subwoofers that are compatible to room size. For example, a 20-Hz signal generates a 55.5-ft sound wave (1100 ft/s ÷ 20 Hz). For this illustration, a 27.5-ft-long room is required. By substituting the longest room dimension for wavelength and manipulating the formula to solve for frequency (frequency = speed of sound ÷ wavelength), the point at which the bass will start to boom can be calculated.

Placing the subwoofer in a corner in the same plane as the front speakers produces the greatest output while maximizing bass depth and power. It also allows the subwoofer to excite all room resonances equally.

Another mechanism of sound does not utilize the inner ear as speakers do. It relies on the tactile (felt) movement of air molecules and other objects against the body. Loudspeakers produce air-transmitted acoustic energy. Tactile movement plays a vital role in experiencing some sounds. For example, when the pilot of a passenger jet throttles up the engines prior to take-off, the people on the plane can easily hear the sound. The other part of the experience is the shaking of the plane at the same time. The passengers can feel the energy of the plane.

Installing tactile transducers is relatively easy. Smaller units are mounted directly to the wooden frames of couches and chairs. Physically mounting these devices to seating surfaces isn't difficult, but it may require some creativity. It may be easiest to bolt a transducer to a piece of plywood, then mount the entire assembly underneath the seating framework using drywall screws.

For best effect, make a direct connection to the seating surface frame. More powerful units can be mounted directly to the floor joists underneath the room. Figure 7.4 shows a typical method.

Powering small transducers with a modest amount of power—25 to 50 W—is normal. However, driving large masses, such as the floor of the media room, requires amplifiers with a minimum of 100 W per transducer. Connecting a tactile transducer to the audio system is similar to adding speakers. Left and right audio signals are run from the amplifier (using outputs such as the tape output jacks) to the transducer amplifier. The amplified signal is routed to the transducers through heavy-gauge (12 to 16 AWG) speaker wire.

Video Components

There are various types of devices commonly used for viewing videos in home theaters. For the present and near future, viewing choices include direct-view sets, rear-projection sets, and front-projection units. Selecting the type of unit for a home theater depends on the room size, lighting in the room, and how the media room is to be used.

The size of the room directly relates to the correct screen size. In general, if the viewing distance is less than 10 ft, a screen size up to 40 in is suitable. Distances of 10 and 20 ft between the viewer and the screen are best served by a screen no less than 40 in and usually no more than 80 in. Distances greater than 20 ft require screens of 100 in or more.

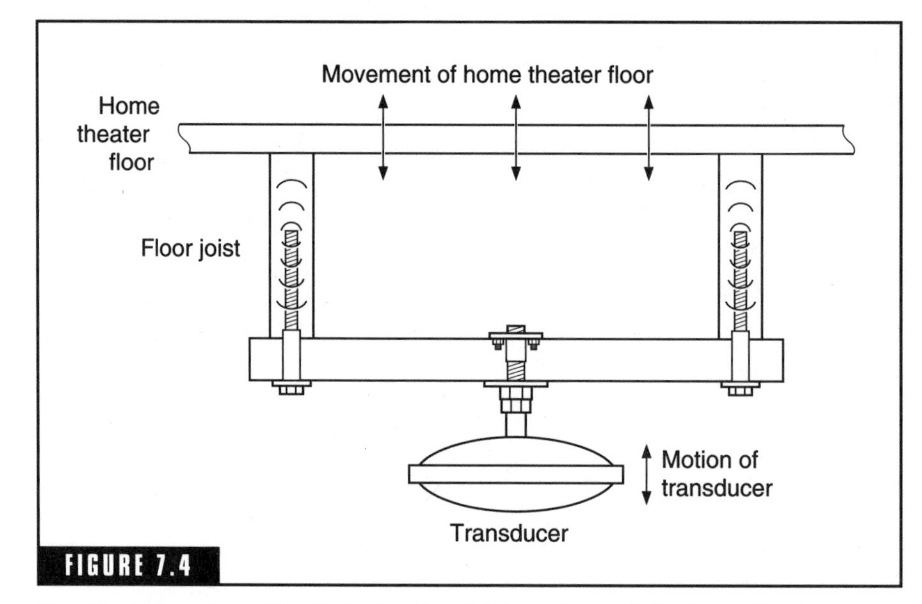

FIGURE 7.4

Mounting a transducer directly to the floor joists underneath the home theater.

The main problem with televisions arises when people watch them in a dark setting. The brightness of the television contrasts too much with the rest of the room; if the set is undersized for the room, this will cause eye discomfort. A bigger television will resolve these issues; however it brings to light another condition. Once the screen reaches 35 in or larger, details such as grain and scan lines become noticeable.

Direct-View Sets

Ranging in size from 13 to 40 in, direct-view sets are what we all refer to as televisions. The screen of a direct-view monitor is actually a cathode-ray tube (CRT). A CRT works by shooting electron beams through a vacuum tube; the beams light up luminous dots, producing the images on the face of the tube. The overall clarity and crispness of a CRT monitor are unsurpassed. Another benefit of direct-view sets is the brightness of the CRT. As long as the sun is not beating directly on the screen, a CRT is easily viewed in any lighting environment. Direct-view sets also have very little need to be adjusted. Most settings are factory-balanced; simply connect to the system and enjoy.

There are some disadvantages to direct-view sets. The first is that as screen size increases, it becomes more curved. Manufacturers are designing televisions with flatter screens; but starting with sets in the 35-in range, the curve is very noticeable and could be very distracting. Another fact that deserves consideration is that the largest direct-view television available is 40 in and is usually more expensive than 50-in rear-projection sets.

A major factor to consider when choosing video equipment is the physical size of the unit. Large or wide-screen CRTs live up to the name—*large.* It is not uncommon for a 36-in monitor to weigh a couple hundred pounds. With the cabinet enclosing the electronics, the big-screen CRTs usually require about 2 ft of depth from the surface of the screen to the rear of the cabinet. Aside from the fact that it takes a good-sized chunk of space out of the media room, some homeowners prefer an inconspicuous home theater installation. Be sure to discuss the options with the homeowners prior to choosing equipment.

Rear-Projection Sets

Rear-projection sets range in size from 40 to 80 in. Large and bulky, these units resemble CRT sets but are usually taller and narrower. Three electron guns (red, green, and blue) project images onto a mirror, which reflects onto the screen. Bigger and flatter than direct-view sets, rear-projection systems are good choices for rooms that are too big for CRTs. Available sizes provide optimal viewing for home theaters.

The only real problems with rear-view sets, besides sharing a size problem with CRTs, is that they are not as bright as direct-view sets. They also require more maintenance than direct-view sets.

Front-Projection Two-Piece Sets

The staple of serious home theaters, front-projection two-piece sets range in size from a low of 50 in to more than 300 in. Also called *LCD* (liquid crystal display) *projectors,* these systems depend on internal lamps to project color images onto a blank wall or a projection screen.

The main advantage is that they provide an opportunity to have a movie theater–sized picture, and they are smaller than almost any other type of set. The size of the image will vary depending on the distance between the projector and the screen.

One of the main drawbacks of the front-projection system is product availability; they are not as widely distributed as rear-projection or direct-view sets. Another problem is that the picture is not very bright. Front-projection or LCD sets, like movie projectors, work best in total darkness. To watch anything on these sets and have a decent picture, there has to be very little light. And finally, unless the unit is ceiling-mounted (Fig. 7.5) or placed high on a shelf, people standing up, stretching, or walking into or out of the room can easily obstruct the projection beam.

Flat-Panel Plasma Displays and Cathode-Ray Tubes

The newest entry into the home viewing market is the flat-panel plasma display. Ultra-thin, lightweight, and appealing to technophiles, these units are actually capable of being mounted on walls, hanging as a picture. Screens vary in size up to a maximum of 55 in.

Plasma displays have front and rear glass plates. Sandwiched between these two plates are thousands of tiny chambers, each containing a mixture of neon and xenon gases. Electrodes send a charge through these chambers to energize the gases. The chambers then glow and create the images seen on the screen.

At present, the quality of plasma screen displays is not high enough to compete with those of other big-screen technologies. Another draw-

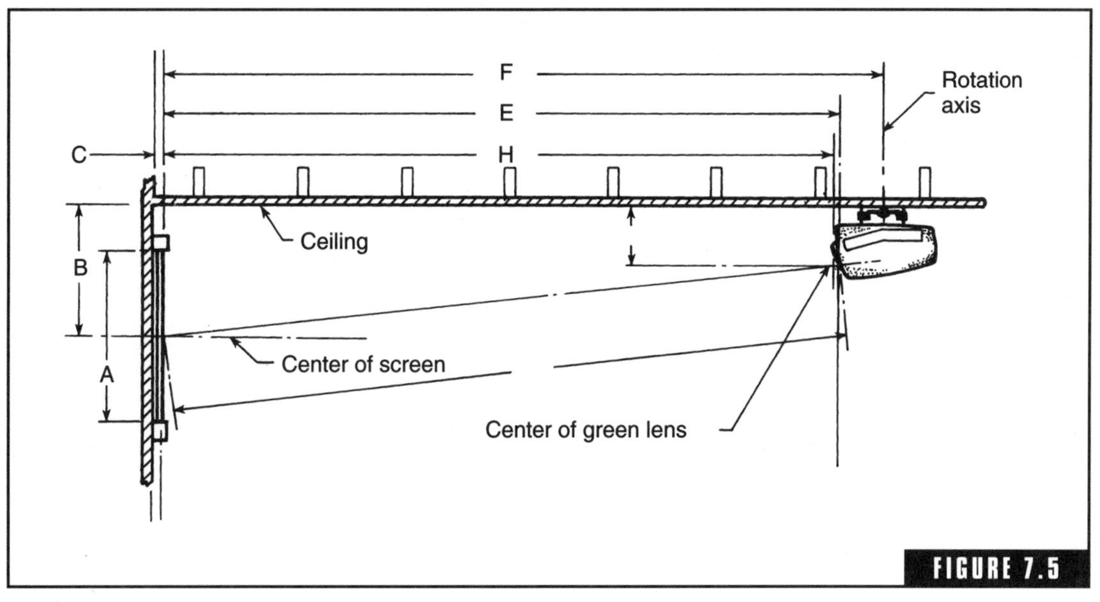

Example of ceiling-mounting a front-projection system.

back is the cost. Flat-panel plasma displays are very pricey, often starting around $10,000. As with most electronics, the future will no doubt see prices fall and quality increase, eventually making plasma technology practical for home theaters.

FLAT CATHODE-RAY TV TUBES

Flat cathode-ray TV tubes are gaining bigger audiences. Television sets are slimming down and gaining intelligence. Many manufacturers are producing entirely new lines of televisions to supply the demand for these new flat TVs.

The market for flat cathode-ray tubes, pioneered by Sony Corporation, is gaining popularity due mainly to their easy-to-see images compared with models with conventional curved tubes. Sharp models are equipped with its Digital Gamma Circuit technology, which brightens dark areas on the screen without changing the brightness of light areas, resulting in a better overall picture.

Toshiba Corporation and Sharp have also made inroads, expanding the variety of flat-display TVs available to consumers. Currently, 23 models are on the market, including 14 Wega series TVs made by

Sony, 4 Face series models released by Toshiba, and 5 TVs in Sharp's Azayaka Flat series.

Sony's Wega is becoming a synonym for flat-display TVs on the strength of its wide product range and early market entry. Toshiba's Face series has substantial functions, including conformity to the text-broadcasting format and the Intertext Vision interactive format. The TVs also feature easy-to-use remote-control units and command displays.

Sony has raised the display density of 32- and 36-in high-definition models by 300 percent to make the coarseness of substandard images inconspicuous on large screens. It has also made its 36-in model as thin as the 32-in model.

HIGH-QUALITY SOUND

All flat-display models of the makers conform to the high-quality sound sources linked to high-definition and digital satellite broadcasting and to digital discs. All feature highly efficient speakers and powerful amplifiers.

Graphics Projectors

When graphics projectors are built, costs often take a back seat to performance. The reason is that graphics projectors are typically among the best products a company can build. Companies frequently view them as technological flagships, so a great deal of research and expense go into their design and assembly. Graphics projectors utilize high-resolution glass or polymer lenses for ultrasharp optical focusing along with regulated high-voltage power supplies for high-contrast pictures without distortion. Magnetic beam focusing with dynamic control is used to provide a complex but superior way to control the electron optics inside the picture tubes for pinpoint sharpness. Another feature of graphics projectors involves digital convergence with microprocessor control. Digital convergence allows for easy and precise registration adjustments. Graphics projectors are known to display the finest video image possible. When coupled with a line doubler, the images projected by graphics projectors are spectacular. There simply is no better video image available.

Due to their high-quality imaging, graphics projectors can display high-resolution computer graphics. Computers are very demanding; the standard VGA, Super VGA, and Macintosh graphics output signals are very high-resolution and require some sophisticated engineering to project cleanly and sharply. Graphics projectors are designed for this demanding work. With the advent of high-definition television (HDTV), display technology will have to be replaced to service the new medium. HDTV has twice the resolution of the current National Television System Committee (NTSC) picture, a wide-screen aspect ratio (16:9 instead of 4:3), multichannel CD-quality audio, and a signal transmission that will be completely digital. Graphics projectors are capable of handling HDTV now.

MULTIPLE ASPECT RATIOS

One of the most exciting things that graphics grade projectors provide is unprecedented flexibility with wide-screen aspect ratio sources. Some of these sources can be displayed now, from letter-boxed movies, and when HDTV fully arrives.

Projection Screens

The screen on which the videos are projected is an integral part of the home theater system. A screen well matched to the system can greatly enhance the overall theater experience; the wrong selection will just as certainly diminish it. Prior to discussing the various types of screens available, it is important to understand some related terminology.

Projection Screen Gain

Gain is the term used in screen reviews and sales literature to describe the amount of light reflecting from a screen surface. Since no actual light is "gained," the term refers to the brightness and directionality characteristics of a screen. The industry refers to flat screens as low-gain screens, while curved screens are known as high-gain screens.

Low-gain screens: Low-gain screens are flat screens like those in movie theaters. Screen gain measured at the front of the screen is moderately bright. These screens are suitable for wide-angle viewing. Recommended room lighting should be as dark as possible.

High-gain screens: High-gain screens, also known as curved screens or rear screens, are very bright at the front of the screen. Viewing is best within a 90° cone (see Fig. 7.6). Subdued room light is best with a high-gain screen.

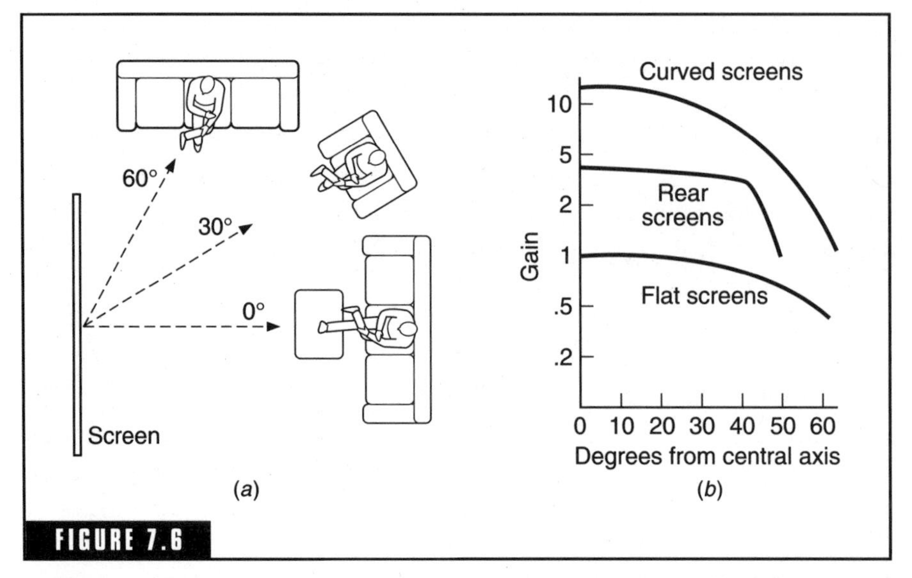

FIGURE 7.6

Optimum viewing angles for high-gain screens. (*a*) Viewing angles; (*b*) Screen gain as a function of viewing angle.

If the homeowners intend to typically use the media room for "special events" and the room can be darkened as a real theater, then a flat, low-gain screen is the proper selection. If the media room is going to be used by the family as a family room with normal lighting most of the time, then a high-gain curved screen (or built-in rear screen) will be the best choice.

Types of Screens

FRONT-PROJECTED CURVED SCREENS

Curved screens weigh very little, are easy to install, and are easily cleaned. These screens are popular with many home theater designers because of the bright, high-contrast pictures they provide. The original projection screens were high-gain curved screens. These screens are large reflective foil surfaces mounted on rigid ABS plastic backings. They are designed to project high-contrast images that are more than 10 times brighter than ordinary matte white screens. One of the basic characteristics of curved screens is directionality. Because they reflect most of the projected image back "on axis," the image is very bright within a 60° viewing cone.

Curved (high-gain) screens were necessary because the video projectors of the time only produced about one-quarter of the lumens (total light) that they do today. In fact, although they still work well in many applications, front-projection systems went through a period of declining sales because they were perceived as "old technology." That trend has reversed due to both the increasing popularity of lower-priced LCD projectors and the inherent ambient light rejection characteristics of curved screens. Generally, when using either a CRT- or LCD-based projector, curved screens are still the best way to create a bright image in rooms with high ambient light.

The reason concerns the curved surface itself. It is designed to direct the major portion of the projected image back into a 90° viewing cone and, as a result, is equally adept at rejecting room lighting off to the sides. In other words, a room designed with lighting and windows off to the sides can result in a bright, high-contrast image on the screen. Figure 7.7 shows an example of both a curved screen and a flat screen for comparison.

FRONT PROJECTION FLAT SCREENS

The other type of screen, and far more universal in home theaters, is the flat variety. These are very popular because they more closely resemble their counterparts in commercial theaters. In some applications, home theater screen material is the same as that used in commercial applications.

The most common flat screen is the manual roll-down type. These screens pull down from their housings for use and are rolled up

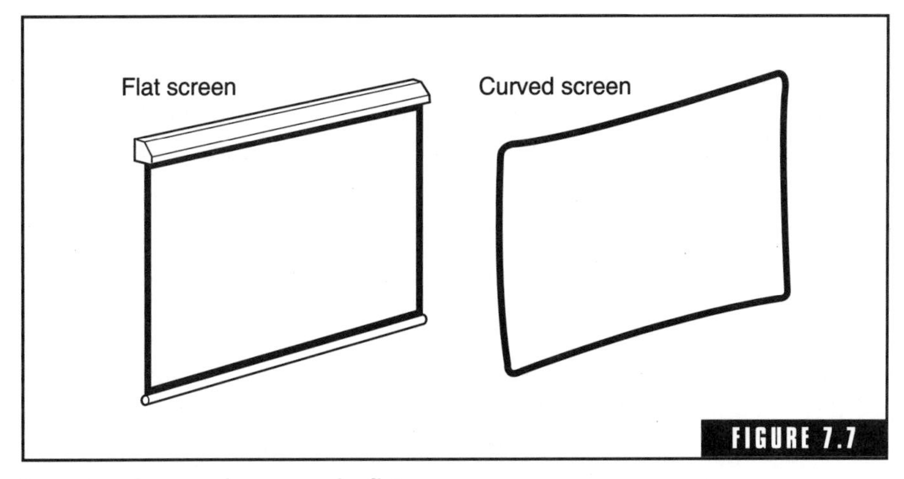

FIGURE 7.7

Examples of a curved screen and a flat screen.

(using tension springs) when not in use. Because it is simple to use and inexpensive, major screen manufacturers offer versions of the manual roll-down screen. A motorized version of the manual roll-down screen is a popular variation. These screens are connected to a 120-V ac power source and can be lowered and raised via a wall switch or remote control. Motorized screens wired to a relay automatically lower the screen when the homeowner activates the system. In existing homes where direct wiring is not an affordable or practical option, battery-operated roll-down screens are available. These battery-operated screens are infrared controlled and charged via the standard ac wall adapter transformer. The advantages are quick installation and ease of operation.

The standard roll-down flat screen comes in a material that is a matte (flat) white with a gain of 1. This material has excellent projection characteristics and a wide viewing angle. For those who want a slightly brighter picture, various manufacturers offer flat screens with higher-gain materials. These surfaces offer brighter images at the expense of viewing angle. Many screen manufacturers offer *tab-tensioned* screens to offset the natural tendency of some screen materials to curl a bit on the edges. These screens utilize external suspension wires to pull the screen surface taut, eliminating wrinkles and edge curl.

STRETCHED FLAT SCREENS

The surface material of these screens is a flexible vinyl sheet that stretches tightly over a black anodized aluminum frame. Stretched taut over their frames, the surfaces of these screens are ultra-flat. And because these screens have such simple construction, they are very affordable and are ideal for dedicated home theaters. Remember that flat screens are very sensitive to ambient light; they tend to look best in dark rooms. Provide control of ambient light via drapes, blinds, and light dimmers.

PERFORATED SCREENS

Perforated screens or acoustically transparent material for home theater use has slightly more than 30,000 perforations per square foot. Each is so small that sound easily travels through the material, yet the holes cannot be seen by the audience. This material permits the front left, front right, and center speakers to be located behind the screen. Commercial movie theaters use this arrangement quite often.

The major advantage of this arrangement is that when sounds emanate from one of these front speakers, the sounds can be localized right near the corresponding on-screen image. This one-to-one correspondence of sound and film image greatly enhances the reality of the presentation. Although perforated screens are the standard in commercial theaters, they are uncommon in home theaters for three reasons. First, it is expensive to perforate thousands of holes in a sheet of vinyl. The machines to perform this process are elaborate and precision-built. Second, perforated screen materials are, by nature, quite fragile and require a fair amount of care during installation. Third, there is a side effect of having many holes in a screen surface: Sound is not the only thing passing through the screen. Light projected on the screen can disappear to the other side. Consequently, the gain of a perforated material is less than the gain of a nonperforated one.

Sizing Home Theater Screens

Homeowners going to the expense of building a home theater will naturally want the largest viewing area possible. However, the largest screen may not be the right choice given the two opposing criteria connected to screen size. An image should be large enough to involve the entire visual field but not so large that it is dim and lacks sharpness.

When screens are sized for use with video-grade projectors, make sure there will be enough room depth that the primary seating is a minimum of 2 screen diagonals from the screen (Fig. 7.8). A system using a line doubling data projection system moves the primary seating from 2 to 1½ screen diagonals from the screen.

Aspect Ratios

The term *wide-screen* most commonly refers to a picture size that is wider than the picture size of U.S. television. Most televisions have a picture ratio (width to height) of 1.33:1. Wide-screen usually ranges from 1.66:1 all the way up to about 2.4:1. The ratio all depends on how the director of the movie, television show, or video wants it to be. The result of having a wider aspect ratio is that there is more information on the right and left sides of the picture.

Most movies are shot with a ratio of at least 1.85:1. During transfer to video or television format, information is lost on the left and right sides

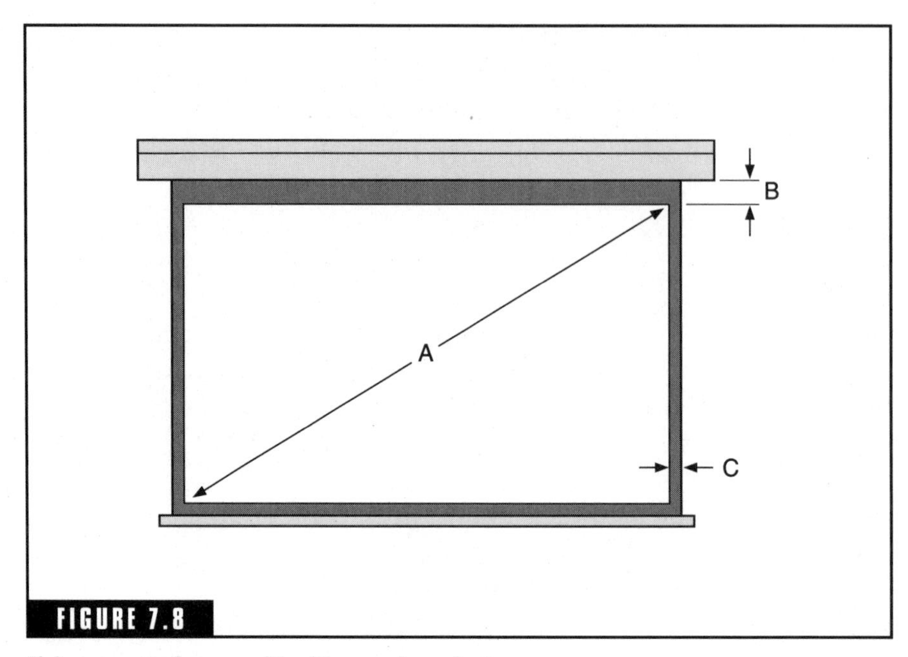

FIGURE 7.8

Sizing screens for use with video-grade projectors.

due to the image expansion done so that the top and bottom of the video match the top and bottom on the television screen. To preserve the ratio that the director intended, movies on television display black bars at the top and bottom of the screen. This process is called *letter boxing.*

The information presented on screen in letter box format is the information the moviemakers meant to be seen. However, most viewers find the bars very annoying. This process of letter boxing reduces the overall resolution and detail of the useful image. Another less-used option that preserves as much detail as possible is *anamorphic squeezing.* All this leads to the advent of wide-screen TV.

Common Aspect Ratios

An aspect ratio of 4:3 (also known as 1.33:1) is the standard aspect ratio in the United States (see Fig. 7.9) and is the standard by which all stations broadcast their signals. Almost all televisions manufactured in the United States for the domestic market have an aspect ratio of 4:3, with the exception of a few wide-screen units. The same can be said for almost all computer monitors. However, wide-screen monitors are now being manufactured as well.

Wide-screen televisions often have an aspect ratio of 16:9 (also known as 1.78:1). In Japan, 16:9 is the standard of high-definition television (HDTV), and it offers a wide-picture compromise between 4:3 and 2.35:1 (see Fig. 7.10).

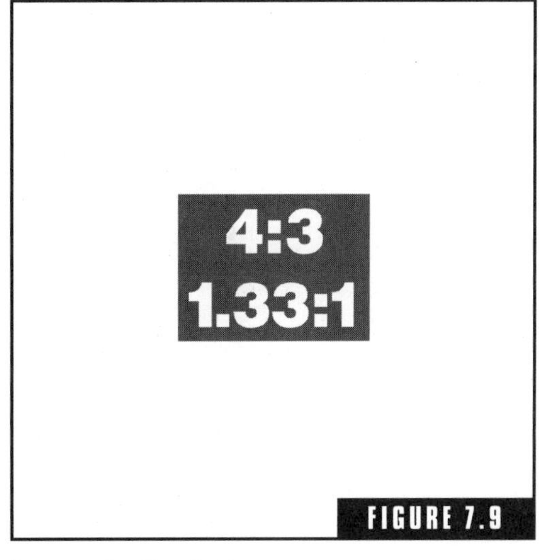

FIGURE 7.9

An aspect ratio of 4:3.

Noise Interference

When cable TV or other externally grounded video sources are connected to the complex audio/video systems found in today's home theaters, it is not uncommon to encounter hum and noise problems. Frequently, the source of a noise problem is the difference in potentials between the home theater ground (provided by the power outlet to which it is connected) and the additional ground paths in the system. This difference results in circulating 60-Hz ground currents, which can cause noise problems. One effect of these 60-Hz ground potential currents is known as *video hum bars.* Hum bars interfere with picture quality, as illustrated in Fig. 7.11.

The difference in audio/video system ground potentials can be caused by a number of factors. One of the most notorious involves the grounding of the cable television coaxial cable where it enters the building. The cable-grounding block needs to be securely grounded to the electric service ground per the National Electrical Code (NEC). If this grounding is poor or nonexistent, external 60-Hz currents can circulate throughout the shielding of coaxial cables. If the system is properly grounded and video hum still exists, it is probably caused by induction from the local house ac wiring.

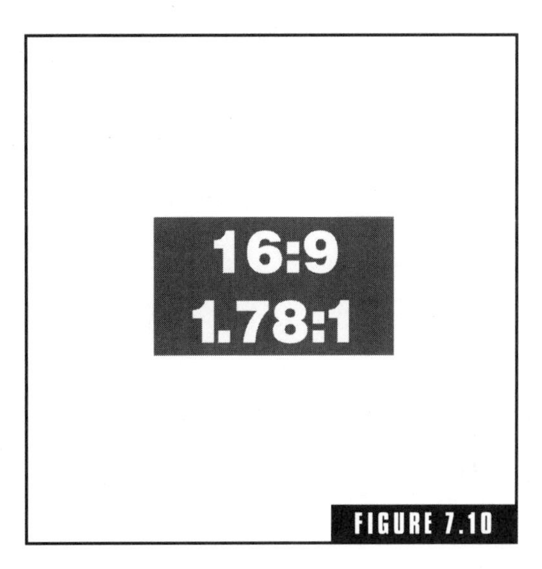

FIGURE 7.10

An aspect ratio of 16:9.

There are a number of ways to eliminate audio and video hum in a home theater system.

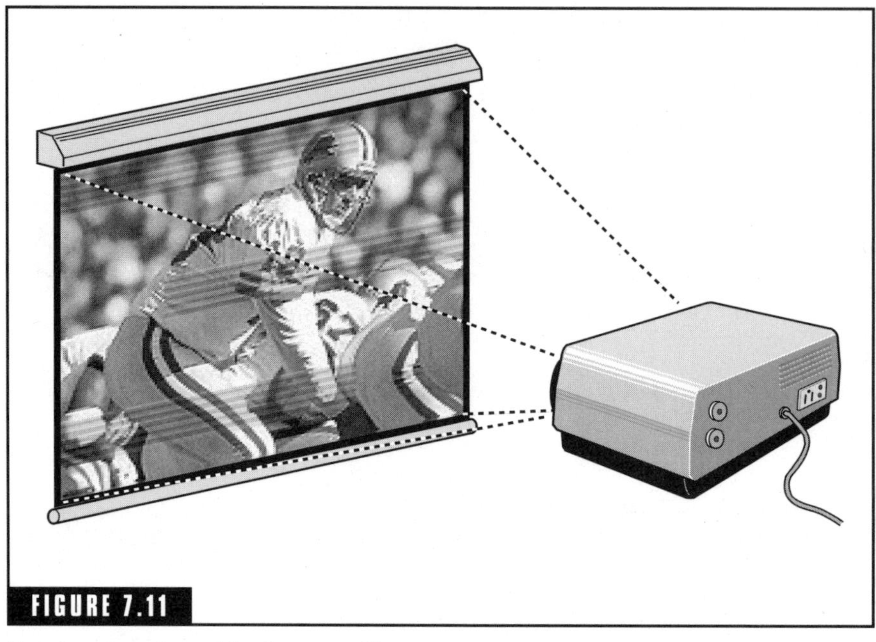

FIGURE 7.11

Hum bars interfere with picture quality.

- Verify that the cable-grounding block is well grounded to the service ground (e.g., not just to a nearby water pipe).
- Plug the entire audio/video system into one outlet. This is often all that is needed to eliminate video hum.
- Disabling the ground to a video display device with a three-pin (hot, neutral, ground) to two-pin (hot, neutral) ac adapter frequently works as well.
- Employ special devices that are designed to stop ground currents from flowing. Begin with the simplest (least expensive) option.

Because audio and video hum is caused by 60-Hz ground currents, often the only thing you need to do is to plug your entire audio/video system into the same outlet so there is no potential difference between grounds. This is easily done with small systems but may be difficult to do with larger home theater systems because they are often distributed throughout the house. The use of power strips which include surge suppressors (Fig. 7.12) provides multiple outlets for devices but draws current from only one source.

Radio Frequency Ground Breaker

Radio frquency (RF) ground breakers are small RF transformers with the primary and secondary physically separated so that the ground path is broken. These inexpensive devices (Fig. 7.13) are often all that is needed to break the path of 60-Hz ground currents flowing through the RF cables. Installation typically places the RF ground breaker in-line with the coaxial cable that feeds into the cable box.

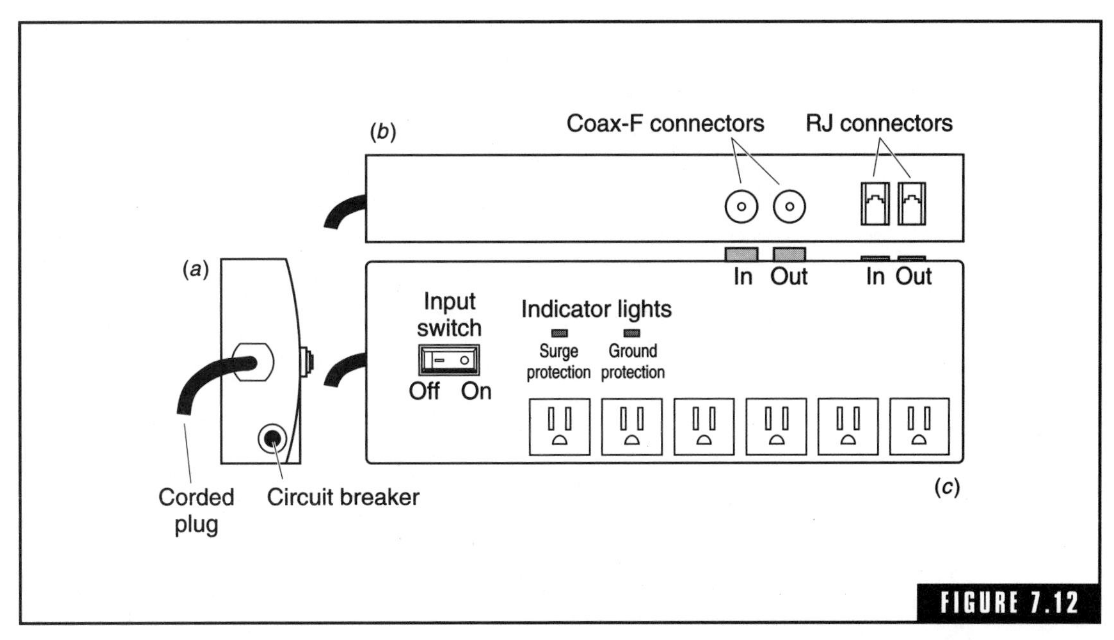

Example of a power strip, which includes a surge suppressor. (*a*) End view; (*b*) side view; (*c*) top view.

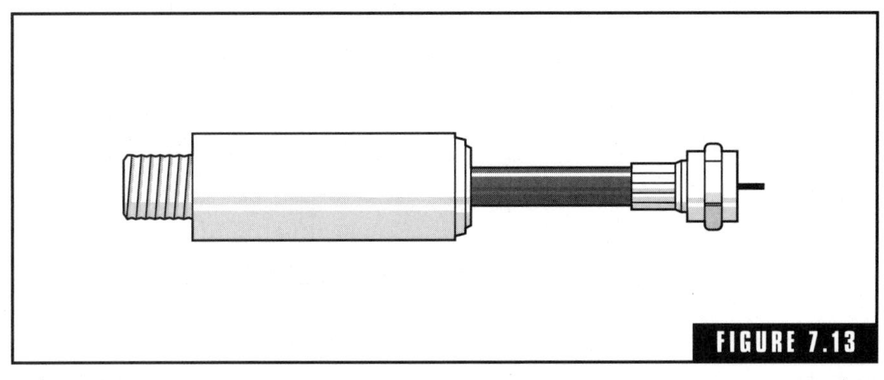

A typical RF ground breaker.

Video Ground Breaker

Video ground breakers are also designed to break the path for 60-Hz ground currents circulating through the video cables. These devices (Fig. 7.14) are wideband transformers with the primary and secondary physically separated so that the ground path is broken. They are typically placed in series with the video cable to the projector.

AC Power Isolation Transformer

AC isolation transformers are common devices in electronic repair shops. Technicians use them to isolate the power on repair benches from physical grounds in the repair shop (pipes, radiators, etc.). This is done primarily for safety reasons.

The same isolation transformers (Fig. 7.15) can be used to electrically isolate audio/video equipment from connected ac power and thus break any ground loops. Installation typically places these transformers in series with the power cord to the projector.

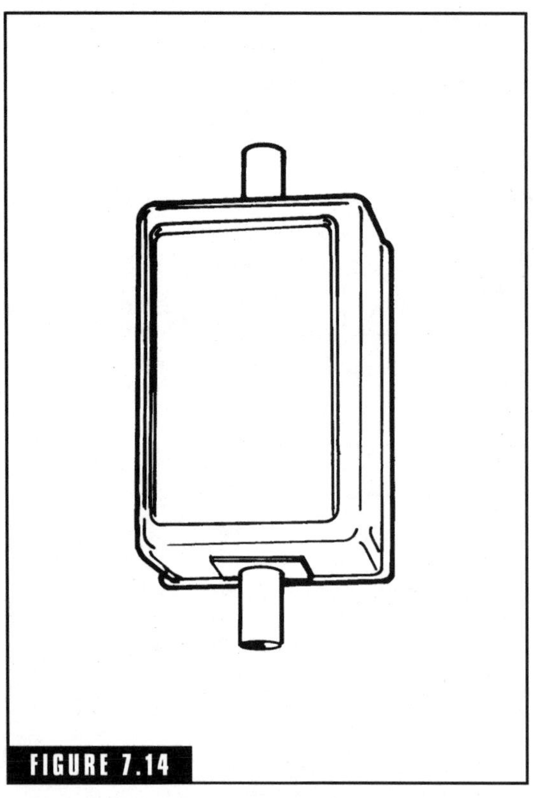

FIGURE 7.14

A typical video ground breaker.

Laserdisc

Not long ago, laserdiscs (LDs) represented the best way to reproduce a film with excellent picture quality. That position has been overtaken by the digital versatile disc (DVD). Normal VHS has only about 210 lines of resolution, LDs have about 425 lines, and the resolution on DVD systems is even higher. LDs have four audio channels, providing sound quality equal to that of audio CDs. There are two types of LD players: single and dual. With a single-side player, the disc must be physically flipped over in order to play the next side. Dual-side players are much more convenient. The laser reader hops over to the other side of the disc, usually in about 9 seconds. Laserdisc players are becoming more scarce, especially single-side players. Projected figures for production of LD media show no increase over the next 5 years. Meanwhile the production of media formatted for DVDs shows a sharp increase for each of the next 5 years.

Most LD movies are presented in the wide-screen or letter box format. The purpose is to transfer the movie and keep it in its original aspect ratio, so you see the movie as the director meant it to be seen.

FIGURE 7.15

Example of an isolation transformer.

Digital Versatile Disc

DVD is a video playback format. DVD used to stand for *digital video disc,* but was changed to digital *versatile* disc because the format is also to be used in the computing arena. VHS or laserdiscs are analog formats while DVD is digital. DVD compresses images to fit on a CD-sized disc. A DVD player decompresses the image and sends it out via different outputs to a display device.

One of the most important capabilities of DVD is that all discs come using Dolby Digital sound. A Dolby Digital decoder is included on all DVD players. However, some DVD players will have six-channel outputs for all the channels, and others will only have two outputs for stereo sound.

Digital Satellite Systems

A *digital satellite system* (DSS) signal begins as a digital broadcast transmitted from a service provider's station to a satellite in orbit. The signal is redirected from the satellite to a small dish (Fig. 7.16) at the homeowner's location. The signal is then routed to a converter box, which changes the digital signal to analog. Once in analog form, the signal continues to the television for viewing.

A typical DSS generally comes with the following equipment:

- An 18-in satellite dish

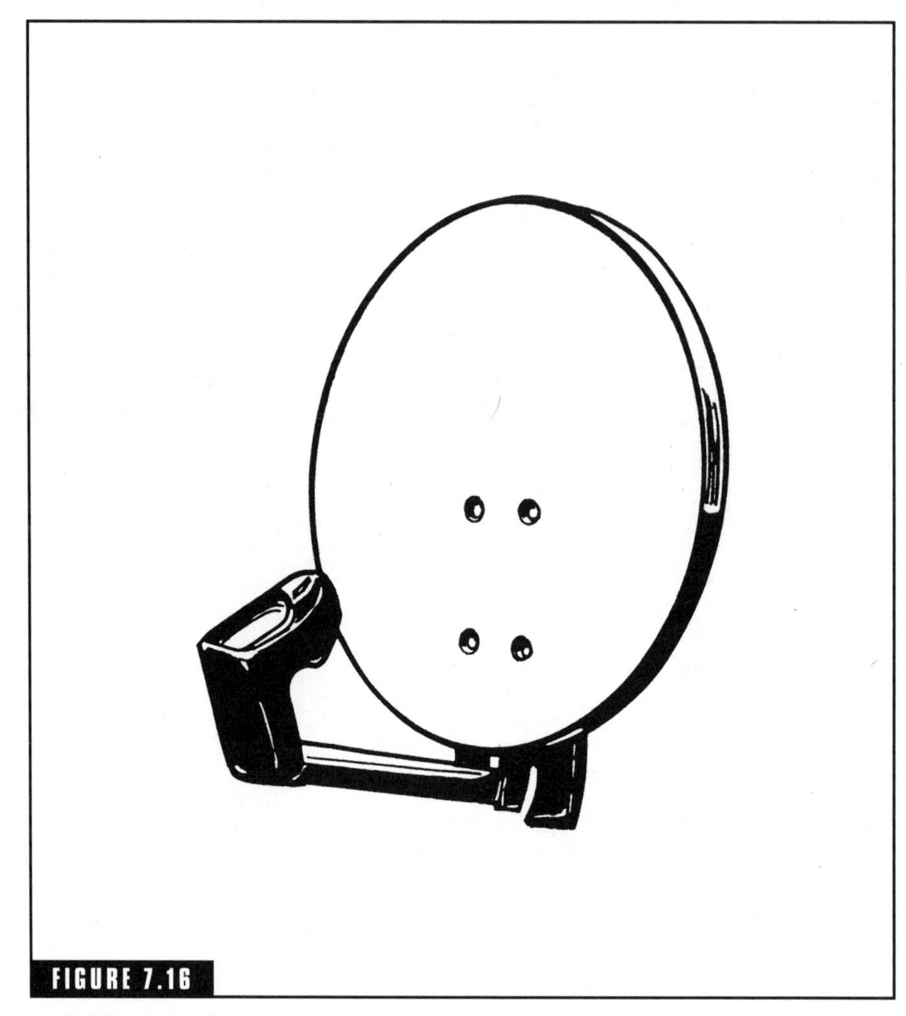

FIGURE 7.16

Typical home satellite dish.

- A mast for mounting the dish
- A converter box
- An access card that provides identification to the service provider
- Miscellaneous cables, usually RCA-type connectors, a coaxial cable connector, and a telephone line
- A remote control with batteries
- A low noise block converter (LNB)

DSS is available in one of two formats. The first is referred to as a *single LNB system.* The LNB is the component of the satellite dish that gathers the signal which has been caught by the dish and sends it to the converter box. A DSS equipped with a single LNB unit will supply signals to only one box. To view two or more DSS programs on two or more televisions or home theater screens or a combination of viewing platforms, a dual LNB system is necessary.

Providers

Hardware manufacturers represent one side of DSS with program providers completing the DSS loop. DIRECTV and USSB are currently the most popular program providers. These companies provide the channel line up that the DSS user chooses from. Channel packages range from a modest number of channels (similar to basic cable) to packages that offer almost 200 channels.

There are some issues for homeowners to consider before committing to DSS. For example, DSS program providers do not broadcast local stations. They do have a network package that includes national networks (ABC, CBS, and NBC). These national networks are tapped out of different cities, such as Chicago or Los Angeles. Local channels must be received from either conventional antenna signals or basic cable television. The cable industry has been successful in protecting its business by having regulations placed that prohibit DSS users from subscribing to network packages if they have had cable service within the last 90 days.

Normally, having a phone line close to the converter box is essential to facilitate services such as pay-per-view and programming such as NFL Sunday Ticket. The service provider downloads any user information needed such as pay-per-view purchases. This also allows the service provider to make sure the box is at the residence where it is supposed to be. Unless you have a phone jack next to where the box will be located, one will have to be run.

Installation

Start by connecting the DSS converter box to the monitor. Use a coaxial cable line to go from an outlet on the box, normally labeled *out to TV,* to the antenna input of the monitor. Turn the monitor channel selector to channel 3 and call up the DSS menu on the monitor's screen. Each menu is slightly different. Check the manufacturer's manual for step-by-step instructions. Once the menu is up, scroll to the section on Dish Pointing or the Installation menu. There you can enter the Zip code to get two pieces of information necessary for the individual installation: the degree of angle at which to set the dish and the azimuth (direction) in which the dish should point.

Using a compass, walk around the outside of the house to locate a position which affords a clear line of sight in the specified azimuth direction—no trees, no chimney, nothing to block signals from reaching the dish. Look for a mounting area as close to the ground as possible, to gain access in case any maintenance is needed. If ground locations are too obtrusive, go to the roof.

Once a suitable mounting location is found, it is time to put up the dish.

- Assemble the dish. Products differ from manufacturer to manufacturer, but all are similar. Do not put the dish on the mast (small bent pole that has the surface mounting plate) yet.
- Set the dish angle degree and tighten the assembly.
- Mount the mast to the house. The mast is made to be mounted on the side of the house in which the dish points. Once you have identified the mounting area, make sure there is a flat spot large enough to mount the dish.
- Mount the mast according to the individual manufacturer's specifications found in the manual.
- Attach the dish.
- Aim the dish in the azimuth direction.
- Run the coaxial cable line inside by drilling at the appropriate spot.
- Ground the dish and cable according to the manual.

Tip: With installations where cable TV lines currently exist, run the wire to the area where the cable line is. Use a diplexor to combine the cable line and the DSS line, allowing them to be transmitted on the same feed. On the other end, near the monitor, another diplexor separates the two signals and makes them ready for connection to the ports labeled *Sat In* and *VHF In*. Once they are attached to the box and monitor, two people can adjust the signal strength by following the process outlined in the manufacturer's manual. When the signal strength is optimized, tighten things down and apply silicon as directed.

Line Doublers

Producing a video image requires sweeping an electron beam across phosphors (chemicals that glow when electrically excited) and changing the intensity of the beam to form a picture with light and dark areas. If you look closely enough at the picture on a television set, you will see the scan lines and colored phosphor stripes that make up the image.

The standard video signal in North America known as *NTSC video* or *composite video* consists of 262.5 horizontal scanning lines per video field. There are two video fields per video frame, equaling 525 lines per frame, and there are 30 video frames scanned per second.

Color video displays use three primary phosphor colors: red, green, and blue. Combining and exciting the three different phosphors can reproduce a complete spectrum of colors. CRT-based projection televisions project the primary colors on top of one another to produce the full spectrum of broadcast colors.

The present 525-line NTSC format originated as a black-and-white standard in the early 1940s; color was added in 1953. The 525-line format was designed so that the average viewer would not see the scan lines making up the image. The format worked well with the picture tubes of the time since they were considerably smaller than what we have today. With today's expanded video images of 100 in or more, scan lines are often very visible.

Line doublers are really just signal processors that take an NTSC video signal (the output of a VCR or tuner) and convert it into a doubled-scan-rate video signal. The result is an image that has 1050 horizontal scan lines versus 525 for NTSC and appears smoother and more film-like. The processing takes place digitally. The analog NTSC video input signal is first demodulated into red, green, and blue signals and then immediately digitized. Next the signal is scan-doubled, motion-corrected, and sharpened, all in the digital domain, then converted back to an analog red, green, blue (RGB) signal. The converted signal now scans at twice the NTSC frequency (31.5 kHz) connected to a data-grade or graphics-grade projector.

A recent development in line doubler technology allows triple and quadruple the number of scan lines. This results in an even smoother picture, but requires a projector that is capable of very high scan rates (such as a graphics projector).

Video Distribution

The whole-house dual coaxial and category 5 wiring discussed earlier in this book should prove sufficient for most video distribution needs. However, there are some fine points to be covered here. When you are planning video distribution, remember that a true home theater experience will occur only in the media room. Signals can be routed throughout the house for viewing and listening to signals generated by equipment in the home theater, but the real theater experience is limited to the media room.

RF Signal Integrity

When video signals are transmitted, it is very important to maintain the signal integrity in order to preserve the picture quality. There are two parts to maintaining the integrity of RF signals. The first part is to keep signals from leaking *out* of the cable. The second part is to keep outside signals from leaking *into* the cable. Both problems are avoided by using good-quality coaxial cable and high-quality connectors.

RF cables are designed on a 75-Ω (ohm) terminated transmissions format. This system is meant to relay RF signals from a source point to a termination point, not from one point to multiple destinations. Routing RF signals to multiple locations cannot be accomplished by wiring all the destinations in parallel.

Another consideration in devising a media distribution system (Fig. 7.17) is control. Most homeowners want to be able to control what they are listening to or what they are watching, even when the equipment generating the signals is in a separate room. Some cities have dual cable systems, so the dual coaxial cable wiring comes in handy in making both A and B cable programming available at all TVs as well as for two-way transmissions.

SPLITTERS

Splitters are used to accommodate multiple demands on a single RF source. Splitters have one input (the 75-Ω load) and two or more out-

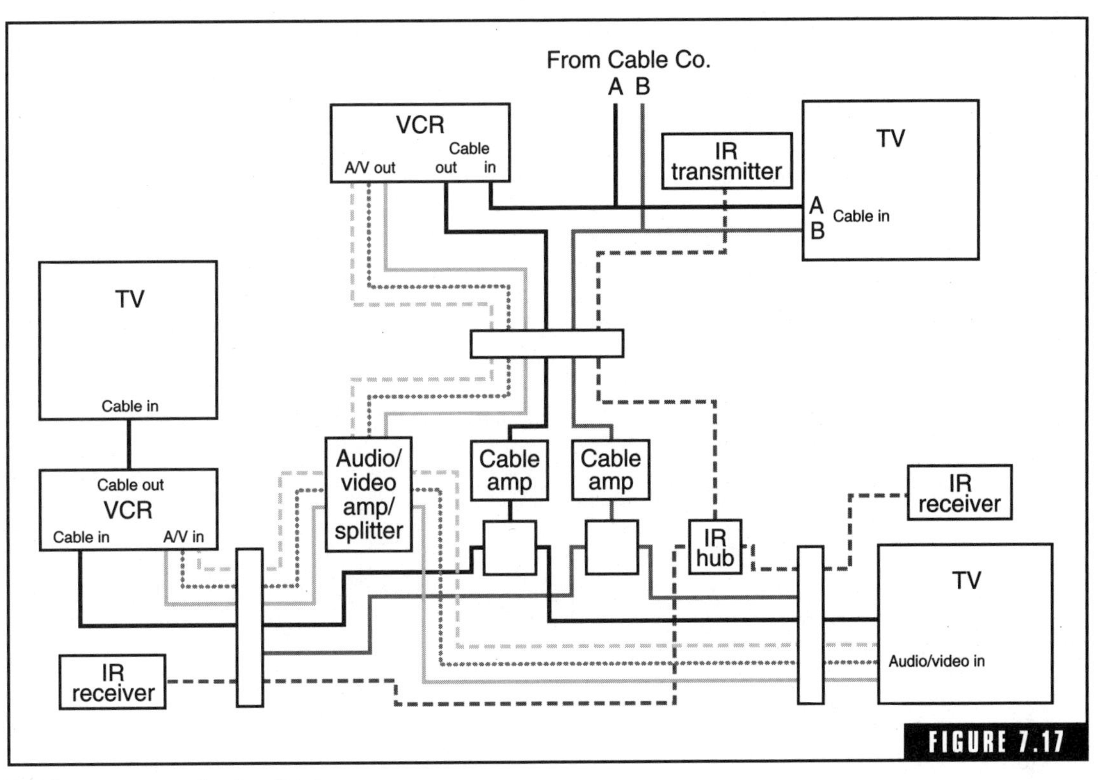

Floor plan of a media distribution system.

puts, each driving a separate 75-Ω load. Essentially they are transformers that split the power in the input signal to multiple outputs, while maintaining the 75-Ω impedance. However, every time the RF signal is split, the signal's strength drastically decreases. An RF signal only has so much power. Logic dictates that splitting this signal in two with a "passive" device will result in two signals that each has—at most—one-half of the original signal's strength. Accordingly, preserving signal strength in multiple distribution systems can best be accomplished by using an amplifier in conjunction with a splitter (Fig. 7.18).

COMBINERS

A combiner simply performs the same job as a splitter, but backward. It combines the channels on two or more separate cables onto one cable. The only drawback is that the cables being combined cannot have any channels in common with each other. Combiners are necessary in video distribution. Connect a modulator to a DSS receiver, and

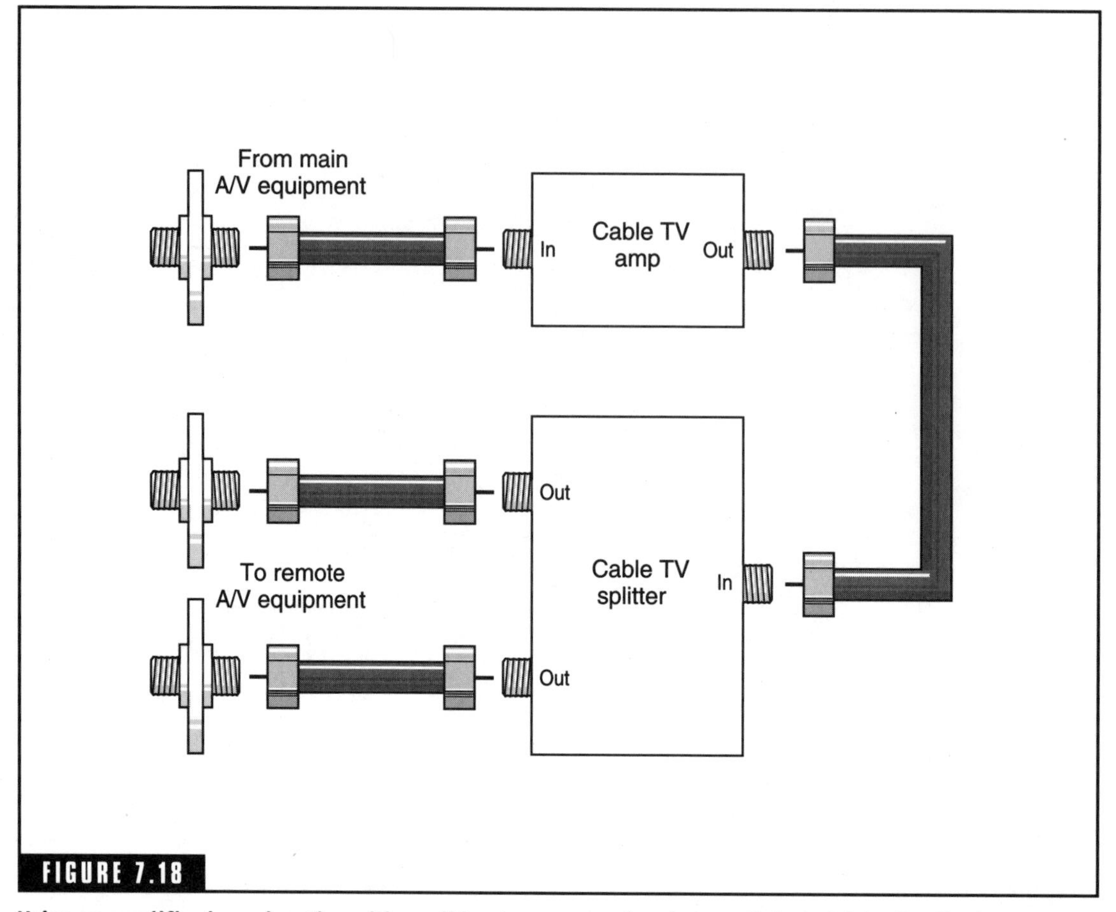

FIGURE 7.18

Using an amplifier in conjunction with a splitter to preserve signal strength in multiple distribution systems.

set the modulator to an unused channel (say, channel 65). Combining this new channel in the wiring closet with the incoming cable TV signal permits any television in the home to receive DSS by changing to channel 65.

RF Signal Loss and Gain

The RF signal loses strength as it passes down the cable and through combiners and splitters. To counter this loss (or *attenuation*), we use RF amplifiers. In the ideal RF distribution system, the signal level at each wall plate should be about the same as the signal level coming in from the cable system or antenna. This ideal is called *unity gain*. By applying a little mathematics and the information in Table 7.2, the approximate losses and gains in a system can be calculated.

RF signal levels are measured as a logarithmic scale of signal relative to 1 millivolt (dBmV). Since decibel values represent power levels and are logarithmic, they can be calculated with simple addition and subtraction. Remember, if the decibel values drop below 0 dB into the negative decibel range, the actual signal information is being lost and no amount of amplification will be able to recover it. In fact, amplifying a signal that is below 0 dB will usually make the picture *worse* since the noise is now being amplified and picked up. Always ensure that signal levels never drop near 0 dB anywhere in the distribution system. This is why the main RF amplifier is usually connected near the input side of the distribution system—so that the signal is boosted early and never drops precariously low.

Audio/video distribution systems route unmodulated video and stereo audio signals from the main location to remote locations. This is most useful when the source is a VCR or laserdisc player, since the RF modulators built into these units normally are not stereo. Each wall panel will have three RCA jacks: one for video and two for audio. One wall panel will be the input to the system, and the rest will be outputs. The input will go through an audio/video amplifying splitter (Fig. 7.19) to keep the signal strong throughout the run.

Interactive TV

Interactive TV allows the viewer to interact with the television set in ways other than simply controlling the channel and the volume and handling videotapes. Typical interactive TV uses include selecting a movie on video from a central bank of films, playing games, voting, or providing other immediate feedback through the television connection. Banking from home and shopping from home are also major applications of interest to consumers and suppliers.

Currently, bringing interactive TV into the home involves adding a special "set-top

TABLE 7.2 Some Rule-of-Thumb Losses for Various Splitters and Cable Lengths

Device	Loss (−dBmV)
Two-way splitter-combiner	4.0
Three-way splitter-combiner	6.5
Four-way splitter-combiner	8.0
Eight-way splitter-combiner	12.0
100-ft RG.6	4.0

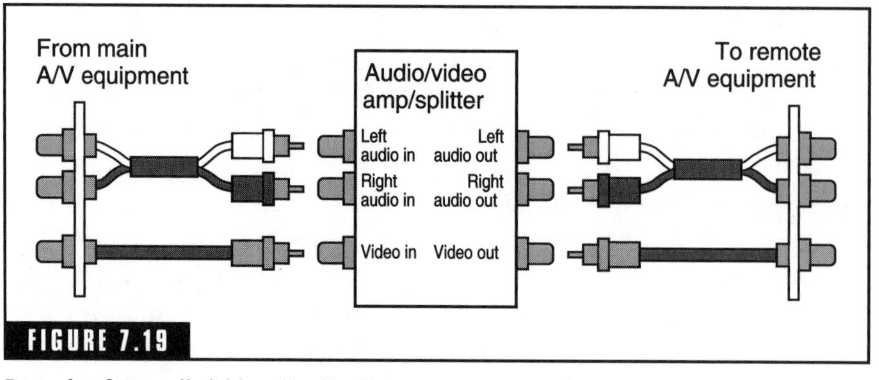

FIGURE 7.19

Example of an audio/video distribution system.

unit" to an existing television set. In addition, other installation and infrastructure arrangements are required, depending on the particular approach. Most services offer special programming, news, and home shopping, and a number offer video-on-demand and home banking. Since interactive TV requires a considerable investment by both the service provider and the consumer, and because interactive applications have been heavily explored over the past 3 years, interactive services are now starting an extensive advertising campaign to lure users.

Interactive TV can be provided through a cable modem, which extends the TV cable bandwidth to the television set or personal computer and is expected to make high-speed Internet and Web access possible. The WebTV Plus system, a next-generation replacement for the original set-top box, is based on an innovative chip that combines the capabilities of a TV tuner, cable modem, and a high-speed PC modem into one unit. This chip enables the new WebTV Plus system to intercept cable TV signals and integrate Web-based information. Information from the Web is retrieved over the system's modem or directly from TV signals through a 1 megabyte per second (Mbyte/s) vertical blanking interval. For example, specific program-related information (TV crossover links), or entire Web pages, can be presented as picture-in-picture, even on TVs that otherwise don't have that capability. The WebTV Plus system contains a 1-Gbyte hard drive for caching previously downloaded information and a 56K modem (the Rockwell K56flex variant). It also contains a standard parallel port that will drive typical computer peripherals such as printers.

Not just WebTV is involved in this convergence of technologies. Other companies are producing and marketing their own versions of *interactive television* (*ITV*). NetChannel, for example, is planning to personalize television using Internet resources by delivering an enhanced television experience, making the Internet personally relevant to individual lifestyles. The NetChannel system uses the Internet as a resource, rather than a destination. It focuses on what is believed to be easier-to-use e-mail and interactive TV program guides enhanced by an "agent" that learns viewer preferences. Their service will become available through set-top boxes, Internet-enabled TVs, and other Internet appliances. Other potential competitors planning to meld the Web and television include such high-profile companies as Sun with its product Diba NetTV.

Interfacing

Combining an intelligent home network with a home theater allows systems and devices to share resources. There are many interfaces available to perform specific tasks; some are very simple and others are somewhat complex. For the sake of example, this section will focus on a wide array of functions possible when home automation connects with home theater. An important feature of a home-network interface device is that it allows connection between the personal computer and a variety of consumer electronic products and home appliances. It needs to support several different physical media and communications protocols in order to mix and match products from various manufacturers. This type of home network integrates lighting, security, comfort, entertainment, and virtually any other electronic products into a unified home system.

Configured as a bridge device, an interface allows other computer-controlled equipment to be integrated into the home network. Interfaces can enable any device with an RS-232 serial port, such as higher-end CD changers, to be added to the network. PC requirements for interfacing include a 486 or higher central processing unit (CPU), Windows 95 operating system, an available universal serial bus (USB), or an RS-232 serial port and the compatible software.

Control

Some of the high-end interface devices send and receive touch-tone signals for optimum control by telephone. They respond to ring, off-hook, and caller ID signals, allowing unlimited communication possibilities. Any combination of telephone signals can be used to trigger events. For example, the system can be programmed so that the television or stereo automatically mutes when the telephone rings and resumes volume when the telephone is hung up. In conjunction with caller ID, incoming calls from friends and family are identified and announced by name through loudspeakers. An interface with a built-in intercom function allows phone-to-phone communication and paging throughout the home via dedicated speakers and/or the audio system. Calls can be placed on hold, announced by paging, and transferred to other extensions. A line-level input allows connection to a microphone preamplifier (or other audio source) for monitoring audio via telephone.

Audio/video equipment is controlled by time, X-10, infrared (IR), telephone, analog, or digital input. Custom programming, known as *macros,* can be used to turn on power, select sources (AM/FM, TV, VCR, CD, etc.). Macros are also used to switch channels, set volume levels, and even close drapes and dim lights at the touch of a button or at preset times. In addition to issuing IR commands, the system can respond to IR commands. Any IR remote is compatible as a controller. It can be arranged so that pressing the Power button on a television remote also dims the lights and closes the drapes. Unused buttons on remote controls can be assigned specific system functions.

Another form of macro is referred to as *IR sequence.* This feature allows predetermined events to be triggered based on a sequence of IR commands received. For example, pressing the TV mute button three times within 5 seconds may dim the lights, draw the drapes, turn on the audio equipment, and adjust the temperature to a preset level in the media room.

Installation is relatively simple. Although manufacturers will have somewhat different designs, reviewing the operating manual should clear up any questions. These devices are basically plug-and-play units with multiple opto-isolated digital inputs to accommodate motion detectors, security sensors, thermostats, switching devices, and other hardwired devices. Digital inputs can be connected in series with alarm system zones for interfacing automation with security. Digital inputs are rated 4 to 24 V ac/dc. Input status is normally logged to a computer file or to the printer for monitoring.

Single-pole double-throw (SPDT) relays allow connection to security systems, HVAC, speakers, sprinklers, and low-voltage lighting. Relay outputs are rated 1 ampere (1 A) at 28 V ac/dc. Automating home security is covered in detail in Chap. 8.

Notes

Notes

Notes

CHAPTER 8

Home Security

Home automation takes home security to new levels. Security systems range from simple stand-alone components to an elaborate array of subsystems interconnected through the automation controller. Sensors, lights, telephones, doors, windows, alarms, and controllers can be implemented to secure and protect individuals and property.

Security Philosophies

During the preconstruction interview between the builder-installer and the homeowners, questions concerning how the family wants security to be handled need to be asked. Before a decision is made concerning what kind of security system to install, it is critical to ascertain from what the system is protecting the home. Homes in an urban environment with large populations are usually threatened with burglary much more than homes in the country.

If homeowners are conspicuously wealthy, important, or famous, the security system takes on better defined duties. People with known medical problems can benefit from integrating a medical alert function into the home security system. Smoke detectors, fire alarms, and carbon monoxide detectors are also critical components of a fully capable home security system. If the property adjoins a public area, such as a park or golf course, some extraperipheral alerts warning intruders away without disturbing the neighbors or calling the police may be a desired part of the system.

Three basic concepts are predominant in the home security industry: deterrence or prevention, detection, and apprehension. For the purposes of this book, deterrence or prevention and detection will be considered the same category. Electronic security systems do offer a great many benefits beyond protection from intruders. A well-designed system can actually simplify daily life.

Deterrence or Prevention, and Detection

Law enforcement studies have shown that the most effective manner of protecting a home against intruders is to make the property a less inviting target. Because of the nature of the work, intruders prefer not to call attention to themselves. A poorly lighted property makes it easy for someone to pass undetected when approaching the house.

Lighting

Although few of us would enjoy living in or paying to illuminate a house like the Washington Monument, properly placed and controlled lighting is the first line of defense in home security. The security aspect of an extraperimeter network of sensors is that it both alerts the homeowner and scares off intruders. A lifestyle benefit is that it is helpful to the homeowner and guests wandering around the property at night. Motion sensors linked to floodlighting are normally placed around the perimeter of the property in overlapping fields of view. Linking these sensors and lights together provides a curtain of lighting around the property. This system activates the lights in response to an intrusion at night, and it can be arranged to activate an audio alert inside the house.

When you are planning exterior lighting, remember to direct floodlight at the house itself; perimeter lights can blind the view of the building, actually hiding intruders. Lighting directed at the house will spotlight anyone standing next to doors or windows.

Sensors

Passive infrared (PIR) detectors, also called motion sensors, will detect any intruder passing in front of the field of detection. These devices emit no signals of their own, but are sensitive to the body heat emitted by a live person moving nearby. PIR detectors can be used to protect any open space that an intruder would be likely to cross. Simple PIR detectors are known for false alarms and are not recommended for a home with active pets. However, dual-technology sensors, using both PIR and microwave sensors to confirm an intrusion, are more reliable.

Other Deterrents

Speakers and warning messages can be added as well, but may prove a bother in the long run. "Radar"-operated electronic barking dog simulators are effective and have much lower maintenance costs than the real thing. Any of these devices will cause no harm if accidentally activated.

Indoor lighting can easily be connected to perimeter lighting. The benefit of this type of arrangement is the appearance of occupancy. Someone wandering onto the property in the dark trips the motion sensor, which turns on the floodlights outside and turns on interior light(s), giving the impression that someone in the house turned on the outside and inside lights. This configuration is especially useful when homeowners are not at home. Working in conjunction with lights controlled by preset time intervals, the complete lighting scheme gives the house a lived-in look.

Note: As with all the security system components, any single home automation protocol can be used, and in most cases a combination of protocols can be configured.

FENCES, GATES, AND LOCKS

Another form of perimeter security involves making it more difficult for people to physically get onto the property. Fencing the perimeter is frequently the first place to start when you are planning a security system. Many types of fences will serve the purpose; specific material and style are the decision of the homeowner. The gate(s) placed in the fence to permit access come in a variety of sizes and styles. Most motorized gates can be controlled either manually or electrically from the house or car.

CARD READERS

Card readers utilize a magnetically encoded stripe on a plastic card to access the gate. Card readers mounted on the inside and outside of the gate require the card to be passed through the reader, which then signals the gate to open.

KEYPADS

Keypads with numerical codes to access the gate are mounted on either side of the gate.

KEYSWITCH

Specially keyed electronic switches installed on either side of the gate permit entry or exit when activated.

IR AND RF REMOTES

Remote infrared and radio-frequency (RF) controllers, like those used for garage doors, are also popular methods of controlling gates.

Motorized perimeter fence gates usually either swing open and close (Fig. 8.1) or roll back along the fence to permit accesses. A magnetic inductance loop is used in the gate to detect the presence of a vehicle. The magnetic inductance loop is used as a "Hold," interrupting the timed-close feature that prevents the gate from closing on a vehicle. This same component when detecting the presence of a vehicle approaching from inside the fence acts as a free exit loop to open the gate, allowing a vehicle to leave. As a safety precaution, motorized gates include a pneumatic edge. This feature stops and reverses the gate if the edge encounters anything. Photoelectric devices known as photocells can also be used as the free exit loop to open the gate as well as the hold loop to interrupt the timed close.

SOUND DISCRIMINATORS

Also known as glass break detectors, these sensors are designed to detect the sound of breaking glass. Some models mount near the middle of the room; others mount close to or on the protected glass. The effective range of glass break detectors is subject to many variables; an electronic tester, designed for the particular sensor, must be used to verify protection. A quality sound discriminator constantly monitors the sounds within its range (normally around 20 to 25 ft) for specific aggressive sounds. For example, the sound of the glass actually breaking will definitely trigger the alarm. Some systems will actuate if an

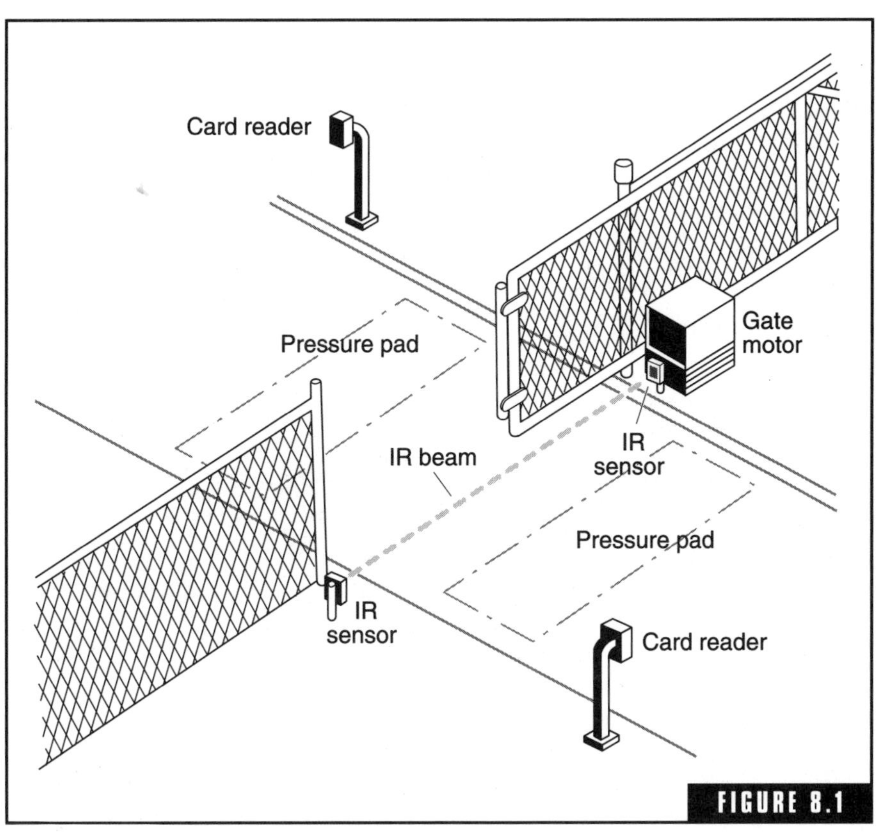

Example of a motorized perimeter fence gate.

object strikes the glass, even without breaking the pane. False alarms are prevented because these systems are programmed to tell the difference between a person knocking on the glass and an implement being used to break the glass.

Since pets and homeowners would normally trip PIR sensors, the sensors are usually put on an "interior only" input on the alarm panel and are armed only when the occupants are away. Sound discriminators are usually found on 24-hour zones.

Closed-Circuit Television (CCTV)

A true marriage between home automation and home security exists with closed-circuit TV. Advances in technology have produced excellent-quality, low-cost cameras for home security purposes. These cameras come in a multitude of varieties; some are specifically designed for outdoor use, while others are intended for use inside.

Using category 5 wiring, these cameras can easily be connected to the home automation system. CCTV cameras can be mounted almost anywhere. Some applications include positioning a camera near the front, back, or side doors, replacing the standard peephole as a means of seeing who is ringing the door bell. Some installations have the cameras mounted on nearby trees, presenting the viewer with wide-angle views of large sections of the house and yard. Remember, for exterior uses, always specify a camera with a weatherproof housing.

CCTV cameras tied into the home automation system can route images to television screens for immediate viewing, or the images can be recorded on a VCR for review later. Homeowners have been employing the later alternative to monitor the behavior of nannies and babysitters. A CCTV camera mounted in the nursery allows homeowners to check on young children without leaving the comfort of the couch.

Doors and Windows

The usual way to protect doors and windows is to install a magnetic reed switch, or contact. Contacts consist of two pieces, the switch itself, and a small magnet concealed in the moving portion of the door or window (Fig. 8.2). When the two are separated by more than a fraction of an inch, an electric circuit is broken and the alarm sounds. The switches themselves either are wired directly to the alarm panel or may be small radio-frequency transmitters that signal the panel remotely.

Obviously, if the window is left open, the switches are ineffective. Alarm screens, which have wires concealed in them and integral contacts, may be used where it is desirable to arm the system with a window open. Any attempt to cut or remove the screen will sound an alarm. Screens are more expensive, but very effective.

Window Installation

The simplest window to protect is a basic, wooden, double-hung window. A window of this kind in a relatively modern home made with Western (platform) framing is relatively easy to protect, especially if the basement beneath it is either unfinished or has a "drop ceiling."

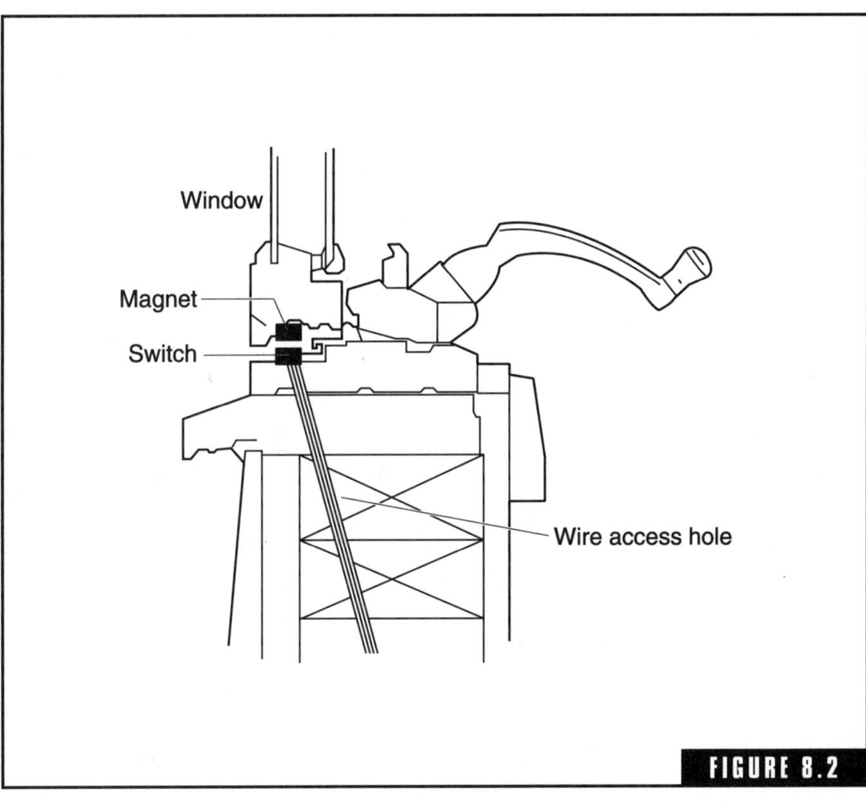

Typical magnetic security sensors.

First, check the wall beneath the window. If there's an electric outlet, stay a few inches to the left or right of it.

Slide the lower sash up, and check the bottom of the sash for any insulation material, such as a rubber strip, which may require the switch to be offset. Also, don't drill closer than 2 in from either side of the sash, as there is sometimes a metal flange in the sash, which can be damaged by the drill. Use a $^3/_8$-in by 6-in (or longer) drill bit, to make a hole in the window sill. Make the hole perpendicular to the sill rather than straight down. Thus when the magnetic contact is inserted, it will fit smoothly and flush with the sill. Drill through the sill, past the shim space, and through the horizontal brace (usually one or two 2 × 4s). As soon as you reach the hollow space in the wall, remove the drill bit from the wall.

Next, insert a $^3/_8$-in by 54-in flexible shaft bit in the drill chuck. The bit is about 4 to 6 in long and has a spring steel shaft welded to it.

There is a bell hanger's hole in the flute of the business end of the drill. There's another similar hole near the tail end of the steel shaft. Hold the drill with the bit touching the floor in front of the window. Place a bit of electrical tape around the shaft to mark the height of the windowsill above the floor. The inside of the wall will be a little shorter than this since the sole plate is at the bottom.

Slip the bit into the newly drilled hole and turn the drill slowly just a few times while working the bit down through the hole. Once the bit is into the hollow part of the wall, just shove it down until the tape mark is close to the windowsill. You don't want to spin the bit yet, or else it may grab the insulation. Wiggle the bit side to side and back and forth a little, to push any insulation away from the fluted part. Now drill through the bottom plate and the subfloor beneath it. The bit should be visible in the basement.

In the basement, a coworker strips off the outer jacket from about 1 ft of 22/4 solid wire and snips off three of the four conductors. The remaining conductor is threaded through the little hole in the flute. Once the wire is threaded, pull the bit back up through the hole. Now gently pull the bit up without turning the motor until the bit catches on the wood. Put the drill in reverse, and very slowly turn it just a few times around as the bit works back through the bottom plate. As soon as it moves freely, stop turning. Pull the bit up again until it catches on the brace just below the window; slowly turn the drill in reverse just a few turns while pulling it up through the window sill.

Pull enough wire out of the sill to work with, clip it off the bit, and put a loose knot in the wire so it doesn't slip back down. Splice the magnetic contact into the yellow and green wires, put a dab of silicone sealant on it, and gently press it into the hole. Using the short 3/8-in bit, drill a hole in the bottom of the sash directly above the sensor. Put a dab of sealant on the magnet, and press it in place. If an insulation strip on the sash was disturbed, put a dab of sealant on that, too. Let the sealant dry for 1 hour before you close the sash. That's all there is to it.

Exterior Door Installation

Many different sensors can be used for standard wood doors in wood frames. For the purposes of this example, we assume 3/8-in cylindrical magnetic contacts. These contacts have a pair of small wires protruding from the back end, which need to be run through to the basement, crawl space, or attic and then connected to the alarm circuit (zone).

Begin by making a $^{3}/_{8}$-in hole in the doorframe, about 1 to 2 in above the threshold. This hole only needs to be about $1^{1}/_{2}$ in deep since the length of the sensor is less than that. Drill the hole into the frame below the latch. The drill will go through the door trim, which is usually less than 1 in thick, and pass through the hollow *shim space*.

The drill bit may have to enter the framing, but it doesn't need to go all the way through it. There are usually wires on the outside of the framing stud running to a light switch by the door. But the wires are *never* run on the inside. You can place a mark on the drill bit if you're not real handy, judging how far you've gone.

Replace the $^{3}/_{8}$-in drill bit with a $^{1}/_{4}$-in by 16-in bit (often called a *feeler bit*). Place the tip of the $^{1}/_{4}$-in bit inside the hole just drilled. Raise the drill up and swing it out a couple of inches. Look down the bit as if it were a gun sight, to be certain it will not come through the door molding. Drill in and down at a sharp angle so that the bit comes out in the cellar between the floor joists.

Pull the bit out and feed a length of 22/4 solid wire into the basement. In the basement, pull the wire over to the control panel location, leaving enough slack to make neat, square corners in the wire. Label the wire to identify its use.

Back upstairs, cut off the wire and solder the two leads from the magnetic contact to the green and yellow conductors coming from the wall. Gently push the excess wire into the hole. Dab a bit of clear silicone sealant on the contact, and press it firmly into the hole. A $^{3}/_{8}$-in hole drilled into the door level with the hole in the doorframe receives the passive component of the system and completes the installation.

Garage Doors

Disconnect the door operator by pulling the emergency release cord. Make sure the door is unlocked to allow the door to move up and down a few inches while installing the switch.

If the roll-up garage door is wooden, place a wide-gap magnetic sensor at the top, running the wires to the control panel along the easiest route. Mount the magnet on the top edge of the door, flush with the outside edge. If the top of the door is made of steel, use longer screws and two to three plastic spacers (usually included with the sensors) to elevate the magnet away from the steel. Mount the magnet with the two arrows pointing toward the switch for maximum gap.

Hold the switch against the door frame (or sheetrock) about $^{1}/_{2}$-in above the magnet, and raise the door a bit to make sure the magnet will not strike the switch. Mark and install the sensor. By placing the magnet $^{1}/_{2}$-in below the sensor, the door needs to travel about $1^{1}/_{2}$ to 2 in to trip the circuit. That takes care of ice buildup, door sag, or other conditions that would cause the door to move.

The garage door in most of our homes is probably the only remote-controlled, motorized door in the house. Because of this uniqueness, garage doors can be equipped to do more than just alert residents of its open or closed status. This segment details a multipart automation idea for electrically operated garage doors using many of the components listed in Table 8.1 for more than one function.

Wire the installation as shown in Fig. 8.3. Mount the normally open switch so that the magnet is aligned with it when the garage door is closed. Most of the modules can be installed together on a power strip which can be fastened to the garage ceiling or to rafters near the opener. The Leviton switch can be installed inside the house in any convenient electric box, and the timer can be plugged into any outlet.

Set both universal modules to "momentary, relay only," and select a house and unit code. Set the Leviton switch to the same house and unit code. Adjust the Powerflash module's input selector to 2 (contact closure), and set the mode to 3 (single unit code on and off). Adjust the house and unit codes to match the appropriate button or light-emitting diode (LED) on the Leviton switch.

The Powerflash module emits a signal in response to the magnetic switch's being activated whenever the garage door is opened. One of the convenience and safety features of this arrangement is that the LED display on the wall switch panel will light up, indicating an open garage door. The LED will turn off only when the door is fully closed. The LEDs on the extra wall switch can be used to monitor other doors or gates by adding magnetic switches and Powerflash modules. Extra buttons on this switch can also be used to close the garage door from inside the home.

When the door is in any position other than fully closed, current flows through the magnetic switch, allowing universal module A to

TABLE 8.1 Various Components Acceptable for Use with Garage Doors

Qty.	Item	Comments
1	X-10 AM466 appliance module	Rated 15 A, 1/3 hp. May need another brand of heavy-duty module if door opener is too large.
2	X-10 UM506 universal modules	
1	Normally open magnetic security switch	*Normally* means when the magnet is next to the switch. May need a wide-gap model.
1	X-10 MT522 minitimer	Not needed with a central controller, i.e., HomeVision, Time Commander, ActiveHome, CP-290, etc.
1	X10 PF284 Powerflash interface	Sometimes called a burglar alarm interface.
1	Leviton 6400 Uni-Base wall-mounted controller	The 6450 plugs into 6400 base.
1	Leviton 6450 modular transmitter keypad	Various models available with different numbers of buttons.

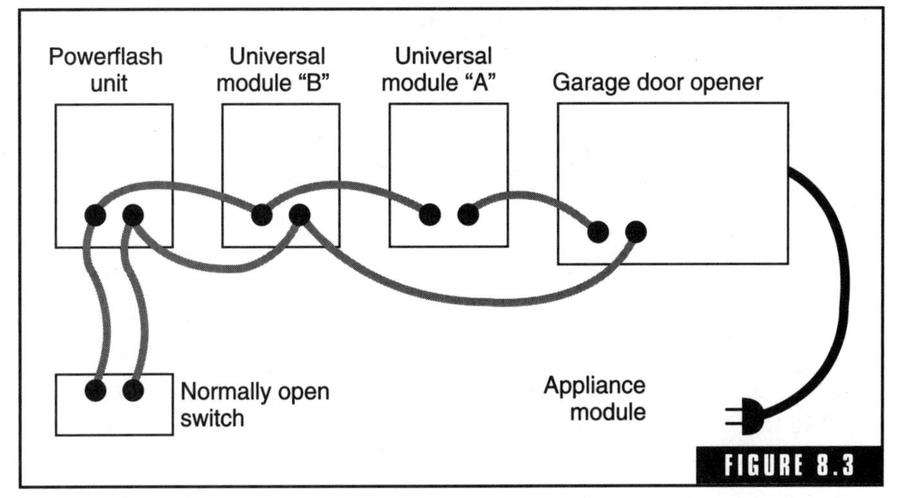

FIGURE 8.3

Tying home automation into an electrically controlled garage door.

close the door. When the door is fully closed, the switch opens and no signal from the universal module can reach the garage door opener, so nothing happens.

Using the minitimer, universal module A can be programmed to activate (closing the garage door) at predetermined times, such as every night at 11 pm. However, if the door was stopped in a partially open position while going down, the timed signal will fully open the door instead of closing it. Sending the signal two or more times with a short delay between will then close the door. This series of signals ensures that the system fully closes the door even if the first signal is blocked by noise on the power line.

Note: Use this automatic closing feature only when the garage door is equipped with a safety beam feature which prevents the door from closing when it comes in contact with an object.

As convenient as it is to have the home automation system close the garage door automatically, X-10 modules are vulnerable to power line noise that can cause errant activation of a module. To remedy this situation, program the system to require two signals before opening the door. To open the garage door, just activate both universal modules. Be sure the universal modules are set to different unit codes or even different house codes for additional security. Because the universal modules are set for momentary operation, there is a limited window of opportunity in which to activate both modules (about 2 seconds).

Wiring the doorbell into the central home automation controller can provide access to the home for latchkey children or for times when homeowners are accidentally locked out. The controller is programmed to accept a code (changable by the homeowner) tapped out on the doorbell. In response, the garage door will open, allowing access to the home.

Many homes are broken into each year by burglars using radio-frequency emitters to open garage doors. With the right configuration, the homeowner can program the garage door not to open (even in response to the homeowner's garage door opener) during preset times.

Plug the garage door opener in through the appliance module. The module can then be scheduled to turn off anytime garage door operation is not desired, such as during vacation periods, while occupants are at work and school, or even during normal sleeping hours. The module can then be scheduled to resume operations again shortly before the residents arrive home.

Fire, Smoke, and Heat Detectors

Every year in the United States, residential fires kill approximately 5000 people, and more than 40,000 people are injured. In addition, home fires do more than $8 billion worth of property damage annually. Many fire-related deaths are caused by the inhalation of smoke and toxic gases, not by burns. Most deaths and injuries occur in fires that happen at night while the victims are asleep. Although it is often not a legal requirement, fire detection equipment is becoming more desired by owners of high-end housing. There are normally three levels of protection: the protection of escape routes, the protection of vulnerable areas, and total coverage. If detectors are used for the protection of property, there are normally two levels of protection: there is first total coverage and then a second configuration where low-risk areas are not covered.

Normally, cabling must be fire-resistant, assuming operation is needed during a fire. This can be achieved by using fire-resistant cables, or by running PVC cables inside conduit, or by burying the cable in walls. Mechanical protection may also be required to protect against impact, abrasion, or attack by rodents.

Any but the smallest building will normally be divided into a number of zones, each of which will have a separate indication on the alarm panel. Often systems will have more than the minimum number of zones, in order to give a more precise location. There are also systems called analog addressable systems in which each device (call point, smoke and heat detector, etc.) can indicate its state and location.

When properly installed and maintained, the home smoke detector is one of the best and least expensive ways to provide early warning when a fire begins. Before the concentration of smoke reaches a dangerous level, or before the fire becomes too intense, the alarm will sound. Smoke detectors save lives, prevent injuries, and minimize property damage. The risk of dying from fire is twice as high in homes that do not have functional detectors.

Smoke detectors work by sensing the rising smoke from a fire and sounding a piercing alarm. There are two types of smoke detectors on the market today. First, ionization chamber detectors use a radioactive source to produce electrically charged molecules (ions) in the air. This sets up an electric current within the detector chamber. When smoke enters the chamber, it attaches itself to the ions and reduces the flow of electric current, thus setting off an alarm. Second, photoelectric smoke detectors (Fig. 8.4) use an internal sensing chamber that includes a light source and a light-sensitive receiver. The chamber is designed so that no ambient light can enter, but smoke can flow through easily. Normally the receiver sees very little light from the light source. Smoke causes the light inside the chamber to scatter. This scattering action causes more light to be seen by the receiver. The detector measures this increase in signal and trips an alarm.

Long-range beam smoke detectors protect a very large area. They can protect an area up to 350 ft by 60 ft. Long-range beam smoke detectors use a separate transmitter and receiver. The transmitter projects an infrared beam, which is measured at the receiver. If the beam becomes obscured by smoke, the receiver signals an alarm. Dust, dirt,

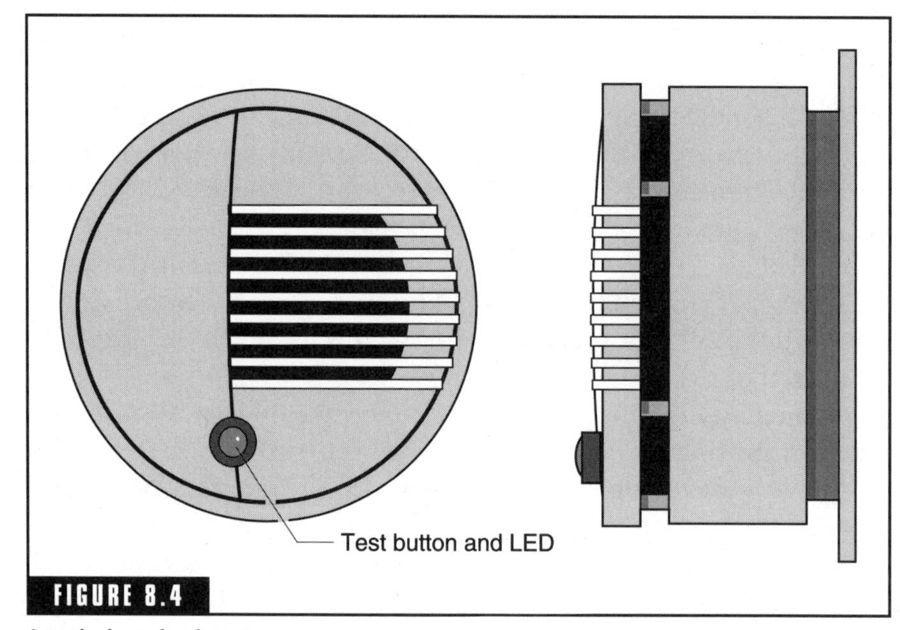

FIGURE 8.4

A typical smoke detector.

or a foreign object blocking the beam causes the receiver to signal a trouble condition.

Smoke detectors also differ by power source. The batteries in battery-powered smoke detectors last approximately 1 year. When the battery begins to lose power and needs to be replaced, the detector will begin to emit beeps every minute or so. Some will keep this up for a week or longer. Smoke detectors that operate on household electric current operate as long as there is current in the circuit to which they are connected. They are, therefore, vulnerable to power failure. Plug-in units must be located near an electric outlet where they will not be unplugged or turned off by a wall switch. They should not get their power from a distant plug by using an extension cord. Always use the hold-in clips to prevent accidental plug removal from the outlet. Ideally smoke detectors should be hardwired and equipped with battery backup.

Each type of detector, if properly installed and maintained, is effective. Since photoelectric detectors react more quickly to smoldering fires and ionization units will respond faster to flaming fires, many installations utilize a combination of detectors. However, because most home fires produce a mixture of smoke types with detectable amounts of large-particle and small-particle smoke early in the fire growth, either an ionization type or a photoelectric type detector will meet most needs.

There should be at least one smoke detector on every floor of the house. Tests conducted by the National Bureau of Standards (NBS) have shown that two detectors, on different levels of a two-story home, are twice as likely to provide enough time for escape as one detector. Although the upstairs detector senses smoke wherever it originates, the downstairs unit will react sooner to a fire that could block escape routes on the first floor.

Several features have been added to smoke detectors over the years. These include detectors that have an escape light, are portable, or transmit their alarm to a central console by radio signal as part of a unified emergency alert system. These can be used with burglar and other warning or detection devices. Electric current detectors with a rechargeable battery for power outages are also available.

System smoke detectors are designed to be connected to a fire alarm system, or as an option to an intrusion alarm system. In a connected system, one smoke alarm being tripped will automatically trip the other detectors throughout the installation. Interconnected smoke detectors are, in fact, utilizing a form of home automation technology. Signals from individual units are carried on the power lines, just as with home automation transmissions. When a single smoke detector is activated, a *message* is carried to all the other smoke detectors, which will be activated by the message.

When connected to the home automation system, the smoke alarms can trigger many responses. The HVAC system can automatically close all the air ducts, preventing smoke from spreading through the house. Lights can be turned on to illuminate the path out of the home. Automatic telephone dialers can call the fire department, while the garage door opens to allow the family to exit and firefighters to enter. After a preset interval allowing time for the family to leave, the lights in the house can be programmed to start flashing, making it easier for fire and rescue personnel to find the site.

System smoke detectors, unlike the battery-operated variety, are connected to the security panel. A monitored security system will automatically notify the central station if smoke detectors are tripped. This allows authorities to respond even if no one is home. It also allows the occupants more time to leave, without concern for calling the fire department. This type of smoke detector is connected with low-voltage wire (Fig. 8.4) and is powered by the security panel. Many local building codes have specific regulations regarding smoke detectors, and may require local ac-powered smoke detectors even if system smoke detectors are installed. Check with local building inspectors to be sure.

Heat detectors are also available, sometimes as part of a smoke detector and sometimes as separate products. These use a special metal that melts or distorts when heat enters the air surrounding it. When built into smoke detectors, these set off the smoke detector's main alarm. Alone, they may sound their own alarm, or they can be tied into a central alarm, as part of a system. Heat detectors add protection, but by themselves they are generally not effective early-warning devices. They must be very close to a fire to be set off. Residential heat detectors are usually used in locations where a smoke detector would be likely to give a false alarm, such as in kitchens, attics, and garages. They are also useful in areas of the home where smoke detectors cannot function because it is too hot or too cold.

Stand-alone heat detectors normally have a preset upper limit for ambient temperatures. Ambient temperatures exceeding these limits will generally activate the alarm. For example, heat detectors with a 135°F fixed temperature should be used in living areas. The temperature is set higher than normal comfort levels to accommodate times when the occupants are away from home with the air conditioning turned off. Heat detectors with higher fixed temperatures, such as 195°F, are designed for use in non-air-conditioned areas (such as attics and garages) where ambient temperatures are routinely high.

Heat detectors are generally equipped with a *rate-of-rise* feature. This rate-of-rise feature constantly measures the temperature of the ambient air. By comparing temperature readings according to a preset time interval, the detector will activate the alarm when a temperature increase greater than or equal to 15°F per minute occurs.

Specialized heat detectors are available for many different applications. With more homes containing home theaters and home offices, multiple cable runs are very common today. A frayed wire or overheated cable can result in a dangerous fire, since it may go undetected until after it has erupted into a full-blown house fire. Linear heat detection systems have become a popular addition to cable trays.

Figure 8.5 illustrates a linear heat detector installed in a sine wave pattern within a cable tray. The detector is run on top of all power and control cables in a tray and is spaced as shown. When additional cables are pulled into the tray, they should also be placed below the detector. The use of messenger wire is optional; however, when not employed, additional fasteners may be required.

CIRCUITS

When you install linear heat detectors, all circuits should be run in series loops. They should not have T or Y branches and should be terminated in an enclosure, which meets specifications supplied by the manufacturer. Class A circuits (four-wire) must leave and return to the main control panel, while class B circuits (two- or four-wire) may be terminated in a remote end-of-line (EOL) resistor or in the main panel.

The maximum length of the heat detection circuit is limited by the capacity of the control panel, which is usually 2500 to 3500 ft (757.6 to 1060.6 m) depending on the model. Copper wire in metallic tubing may be used to reach the areas to be protected, but only the linear heat

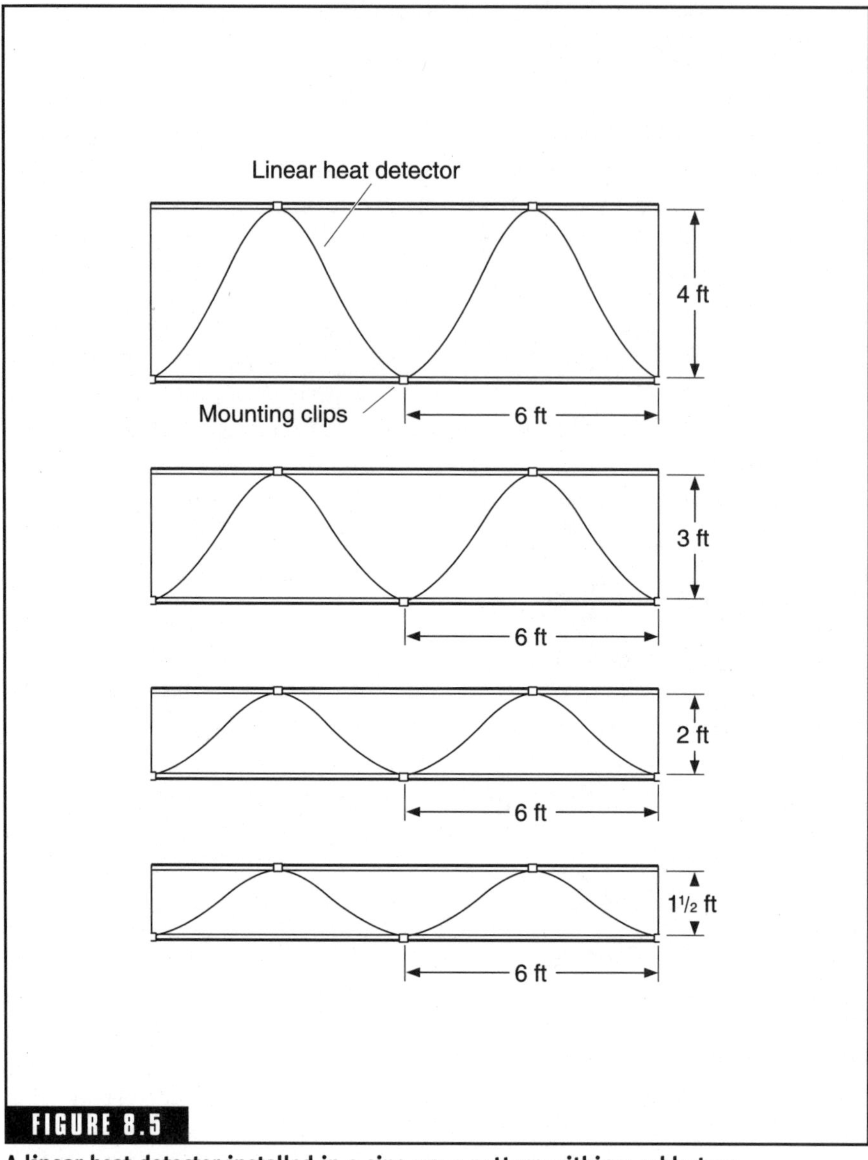

FIGURE 8.5

A linear heat detector installed in a sine wave pattern within a cable tray.

detector should be employed for any part of the circuit which is intended to detect overheating or fire.

There may be parts of a protected area in which the wire in the circuit is not expected to be a detector. Such is the case in locations which have extremely high ambient temperatures or when it is necessary for a circuit to cross another active detection circuit to reach the intended area to be protected. Under these conditions, copper wire should be installed in these limited sections only, and it should be spliced to the linear heat detector in suitable enclosures.

FITTING AND FASTENING

All details of installation should be performed in a neat and workmanlike manner. All bending and fitting of the linear heat detector should be done with the fingers. Pliers or other hard tools should not be used for this purpose. All bends should consist of rounded turns. No 90° bends should be made.

Figure 8.6 illustrates wire straps used in covered cable trays with rolled edges and power distribution apparatus such as transformers and motor control panels. Figure 8.7 illustrates the application of mounting clips on various types of cable trays.

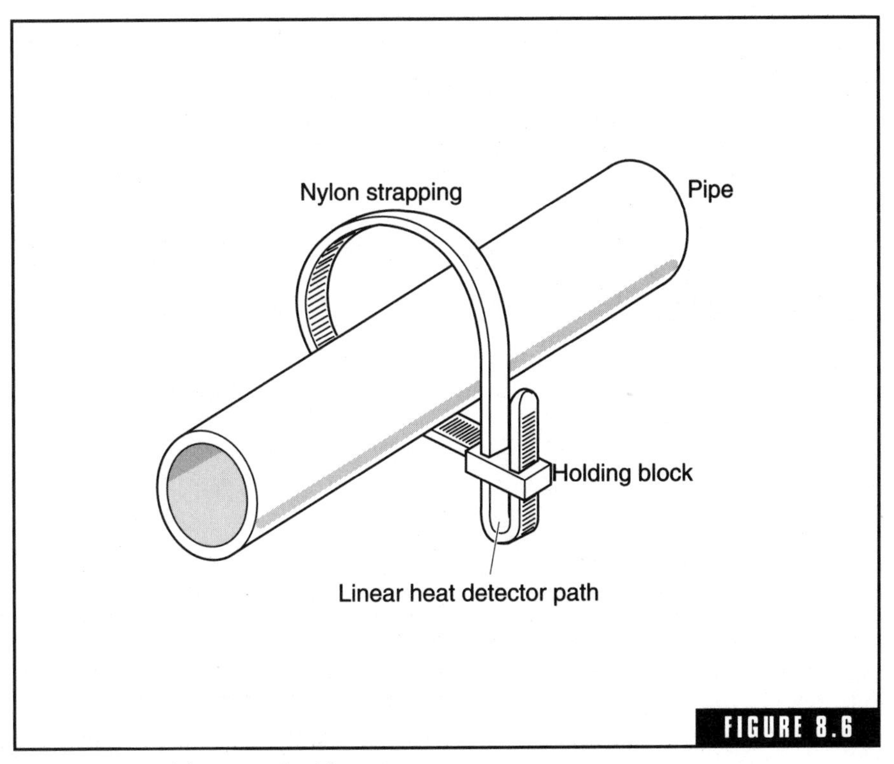

FIGURE 8.6

Wire straps used in covered cable trays.

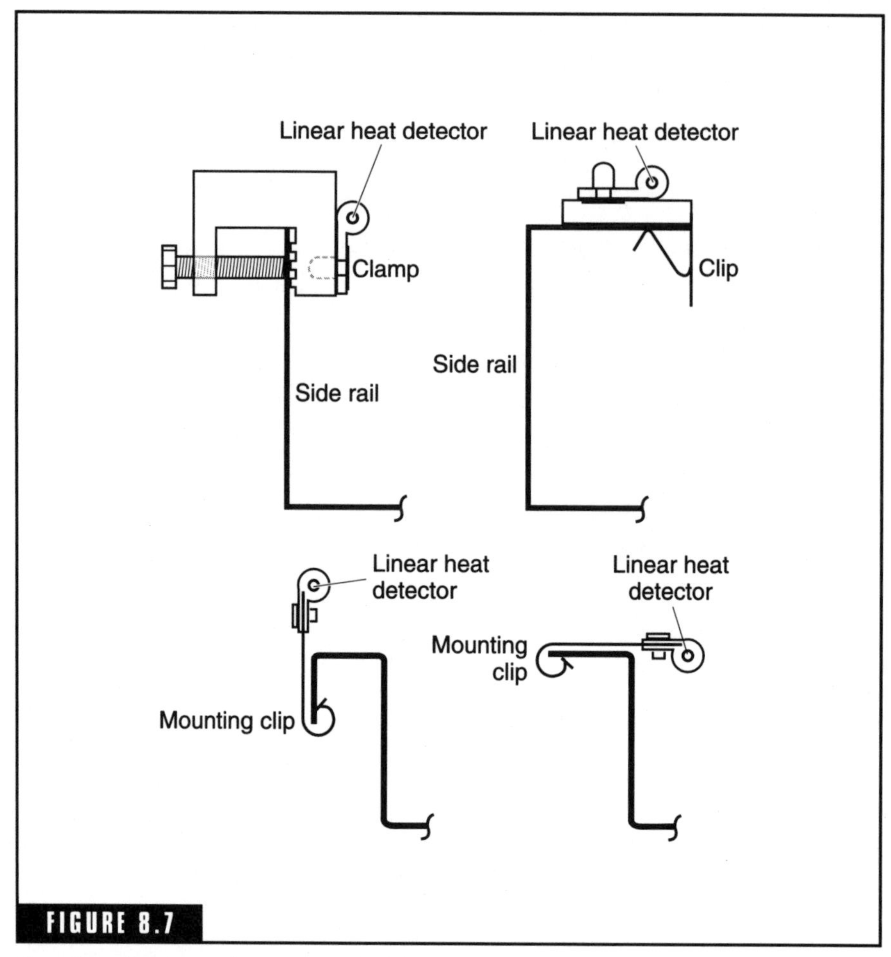

FIGURE 8.7

Mounting clips on various types of cable trays.

MECHANICAL PROTECTION

The linear heat detector should be enclosed in conduit wherever it is connected to manual fire alarm stations, test terminal control panels, or junction boxes. The open ends of metal conduit through which the detector passes need to be provided with bushings. After the control panel is mounted and all conduit and wiring are installed, the panel doors should be left closed. Both ends of all conduit or wireways, which connect to the control, should be completely closed off with duct seal so that no gases or condensation will be able to reach the inside of the control cabinet. Manual stations and end-of-line resistors (not in the control panels) should be installed not more than 6 ft (1.8 m) above floor level.

CONNECTION AND SPLICING

All connections made to terminals are normally made using soft copper, flexible leads or compression terminals. Splices are sometimes necessary; proper splicing is critical to the operation of the system. Some manufacturers supply splicing sleeves or splicing connectors to assist in making effective splices.

Outdoor applications require the use of sealant tape or appropriately rated junction boxes for in-line splices. Exposure to direct sunlight may cause the temperature of the linear heat detector or its mounting surface to exceed the maximum ambient limit and/or the operating temperature rating of the detector. Shielding of the detector may be required in order to reduce the maximum ambient temperature to acceptable limits.

Carbon Monoxide

Carbon monoxide cannot be seen, smelled, or tasted; but at high levels, it can kill a person in minutes. Carbon monoxide (CO) is produced whenever any fuel such as gas, oil, kerosene, wood, or charcoal is burned. If appliances that burn fuel are maintained and used properly, the amount of CO produced is usually not hazardous. However, if appliances are not working properly or are used incorrectly, dangerous levels of carbon monoxide can result.

Hundreds of people die accidentally every year from CO poisoning caused by malfunctioning or improperly used fuel-burning appliances. Even more die from CO produced by idling cars. Fetuses, infants, elderly people, and people with anemia or with a history of heart or respiratory disease can be especially susceptible.

Homeowners are not the only ones at risk. Contractors and installers are often overcome by CO when working in enclosed spaces and rooms where a malfunctioning appliance is operating. Know the symptoms of CO poisoning. At moderate levels, many people experience severe headaches and become dizzy, mentally confused, nauseated, or faint. Eventually, death can result if exposure to these levels persists for a long time. Low levels can cause shortness of breath, mild nausea, and mild headaches, and may have longer-term effects on the general health of those exposed. Since many of these symptoms are similar to those of the flu, food poisoning, or other illnesses, CO poisoning is often not immediately thought to be the cause. Do not ignore symptoms, particularly if more than one person is feeling them. Carbon monoxide poisoning can cause rapid loss of consciousness and death; if symptoms are noticed, act immediately. Get fresh air immediately. Open doors and windows, turn off combustion appliances, and leave the area. Go to an emergency room, and tell the physician you suspect CO poisoning. If CO poisoning has occurred, it can often be diagnosed by a blood test done soon after exposure.

The great danger of CO is its attraction to hemoglobin in the bloodstream. CO is breathed in through the lungs, and it bonds with hemoglobin in the blood, displacing the oxygen, which cells need to function. When CO is present in the air, CO rapidly accumulates in the blood. It will eventually displace enough oxygen in a person's system to cause suffocation from the inside out, resulting in brain damage or death.

The U.S. Consumer Product Safety Commission (CPSC) recommends installing at least one carbon monoxide detector per household, near the sleeping area. A second detector located near the home's heating source adds an extra measure of safety.

Similar to smoke detectors, carbon monoxide detectors are available for homes, garages, and other areas to warn of CO poisoning hazard. The technology of CO detectors is still developing. There are different types of CO detectors on the market, and generally they are not considered as reliable as the smoke detectors found in homes today. Some CO detectors have been laboratory-tested, and their performance varied. Some performed well, others failed to alarm even at very high CO levels, and still others alarmed even at very low levels that don't pose any immediate health risk. In addition, unlike a smoke detector, where you can easily confirm the cause of the alarm, CO is invisible and odorless, so it's harder to tell if an alarm is false or it is a real emergency.

Carbon monoxide detectors not only are available in a variety of shapes and sizes but also are manufactured for various installations. As seen in Fig. 8.8, CO detectors can be hardwired into the ac power lines, plugged directly into an outlet, or battery-operated.

The units best suited for use in a total home system can be hardwired in the same way as smoke detectors are. The units are tied into

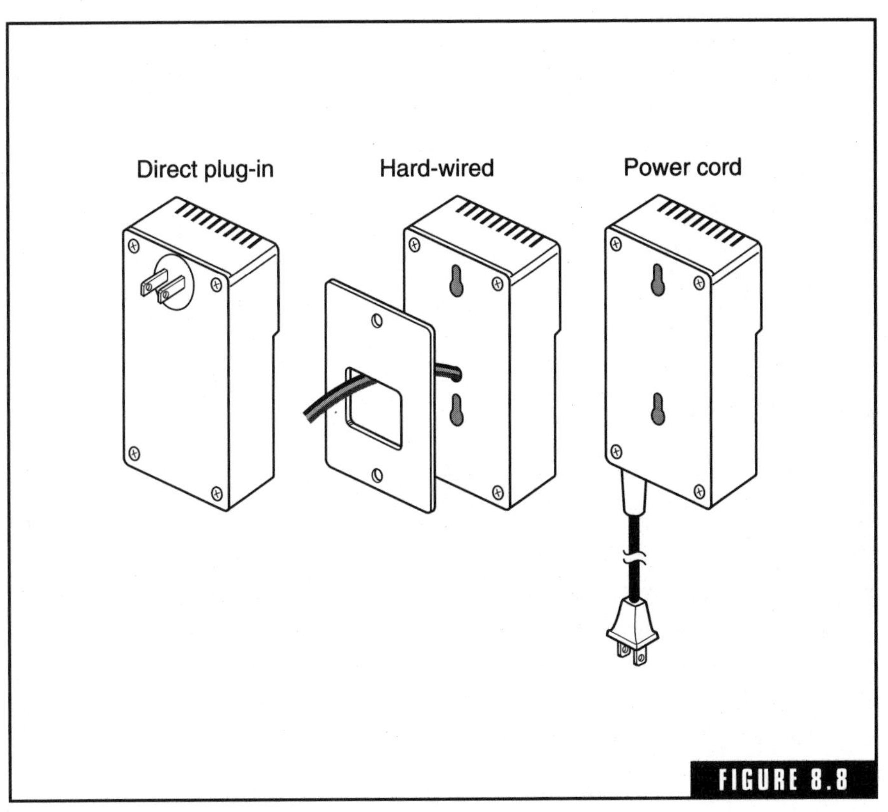

Carbon monoxide detectors can be powered in a variety of ways.

the home automation system in the same manner, allowing automatic dialers, lights, and alarms to function according to programming.

SOUNDING DEVICES

Fire, smoke, heat, and carbon monoxide detectors all share a common feature: They usually sound an audible alarm to alert the occupants of danger. Various alarms are available depending on the requirements of the situation. It is advisable to install different sirens for different detectors to make each alarm easily recognizable.

SELF-CONTAINED SIREN

This type of siren has the electronics necessary to drive the speaker built into the speaker housing. The wires that are attached to this type of siren can be connected directly to the control panel alarm output terminals. If separate speakers are used instead of the self-contained speaker, a separate siren driver will be necessary to provide the electronics to drive the speakers.

WARBLE OR YELP TONE

This tone is created by the changing of the frequency of the tone to produce a rising and falling of pitch. The stereotypical "police car" siren sound best describes this sound. Most alarm installation companies use this tone for burglar alarm sirens.

Automatic Dialers

In order for a security system to communicate with central monitoring stations, fire stations, police, medical personnel, or friends and neighbors, it is necessary to connect the system to the phone line. However, alarm systems require a special type of connection to the telephones, or they simply won't work correctly. Begin by running category 5 low-voltage wire from the alarm system's master control panel to the telephone block (NID). Note which terminal on the protector has the red wire from the house telephones and which has the green wire. Disconnect them and follow these steps:

- Connect the red wire from the box to the red wire from the street (the telephone company line).
- The green wire from the box is connected to the green wire from the street.
- Splice the yellow wire from the box to the red wire from the house telephone jacks.
- Splice the black wire from the box to the green wire from the house telephone jacks.

Back at the box, connect the new wire to the RJ-31X jack as follows:

- Connect the red wire to the terminal that has an existing red wire connection.
- Connect the green wire to the terminal that has an existing green wire connection.

- Connect the yellow wire to the terminal that has a brown wire connected to it.
- Secure the black wire to the terminal where a gray wire is currently connected.

Normally, one more step is necessary. Connect the gray modular cord to the alarm, and plug it into the RJ-31X jack, following the alarm system's schematic drawing for proper positioning of the red, green, brown, and gray wires from the cord. Bend the other pairs of wire out of the way, and cover them with electrical tape so they don't short-circuit anything in the panel. Connect the wires to the correct terminals.

EOL RESISTORS

The abbreviation *EOL* stands for *end-of-line*. The purpose of an EOL resistor (or EOLR) is to supervise a circuit against tampering or accidental short or open circuits. Most modern alarm control panels have the option of supervising any or all burglar alarm loops (zones) with an EOL resistor. There are two types of powered detectors, forms A and C. Form C has a C (common) terminal, an NO (normally open) terminal, and an NC (normally closed) terminal. Form A output has only the C and NC terminals.

Form C

- Connect the red lead to the positive (+) terminal.
- Attach the black lead to the negative (−) terminal.
- Secure the green lead to the C terminal.
- Pair up the yellow lead with the NO terminal.
- Attach the EOLR across the C and NO terminals.

Form A

- Connect the red lead to the positive terminal.
- Attach the black lead to the negative terminal.
- Run the green lead to the C terminal.

- Connect one leg of the EOLR to the NC terminal.
- Secure the yellow lead to the other leg of the EOLR.

Smoke detectors are wired as follows:

Four-Wire

- Connect the red lead to the positive terminal.
- Attach the black lead to the negative terminal.
- Connect the green lead to the C terminal.
- Secure the yellow lead to the NO terminal.
- Secure the EOLR across the C and NO terminals.

Two-Wire

- As usual, connect the red lead to the positive terminal.
- Attach the black lead to the negative terminal.
- And secure the EOLR across the positive and negative terminals.

Magnetic sensors (contacts) are made in three ways: NC, NO, and SPDT. SPDT is another way of saying form C. The NC sensors are closed when the door or window is closed. They open when the door is opened. The NO sensors are just the opposite—they open when the door is closed and close when the door is opened.

NC

- The red and black wires are not used.
- Yellow wire goes to one side of the sensor.
- One side of EOLR goes to the other side of the sensor.
- Green wire goes to other side of the EOLR.

NO

- Red and black wires are not used.
- Yellow wire goes to one side of the sensor.

- Green wire goes to the other side of the sensor.
- EOLR runs across (parallel to) the sensor.

SPDT

- Red and black wires are not used.
- Yellow wire goes to the C side of the sensor.
- Green wire goes to the NO side of the sensor.
- The EOLR goes across (parallel to) C and NO.

WIRE TYPES

Many manufacturers specify similar wire for installation of the security systems and components discussed in this chapter. Some of the most frequently recommended wire types are as follows:

- 22 AWG two-conductor wire: includes outer insulation covering the two insulated conductors. Use to connect external contacts to transmitters. Also use to connect inside sounding devices to control panel terminals.
- 22 AWG four-conductor wire: includes outer insulation covering the four insulated conductors. Use to connect consoles to control panel terminals. Also use to connect the RJ-31X to the phone company's interface box (suggest gray color).
- 18 AWG two-conductor wire: outer insulation covering the two insulated conductors. Use to connect external power supply to control panel terminals and outside sounding devices to control panel terminals.
- 14 AWG single-conductor wire: used to ground the security system.

Apprehension

As the name implies, the security philosophy of apprehension involves taking measures to catch the would-be intruder. All the devices mentioned in this chapter can be used for this purpose, just as they are useful in detection and prevention. The main difference in a system designed for apprehension revolves on alerting the intruder that his or her presence has been detected.

In a deterrent-based security system, intruders are immediately made aware of the fact that they have been detected. The idea is, once detected, the intruder will flee the area to avoid capture. Apprehension-based systems do not sound audio alarms or turn on lights. The system may sound a silent alarm, which registers at the remote monitoring station or police station. Another alternative is the automatic telephone dialer, which calls the police or security force and relays a prerecorded message notifying authorities of the need to respond to the situation. The idea here is for the authorities to arrive and capture the criminal in the act.

Response Time

Monitored alarm systems use a digital communicator, which seizes a telephone line in the event of an intrusion and signals a computer at the monitoring center or central station. An operator receives the alarm and follows specific instructions to notify the proper authorities. In most cases, the operator will verify the alarm by calling the home, before notifying police.

Monitoring centers are usually high-security operations, operating 24 hours per day. These stations are equipped with backup power supplies, telephone lines, and computers. Some use long-distance 800 lines and redundant sites, to ensure operation even after a local disaster. Some monitored systems are equipped for two-way voice communication between the home and the central station. This allows the operator to better assess the threat before calling for a dispatch.

Response times of local police authorities vary greatly from town to town. Because burglar alarms have historically experienced a high incidence of false alarms, many police departments will respond to an electronic alarm only after it has been verified. In major cities, response to an unverified alarm may be 1 hour or more. In those cases, two-way voice verification or use of a private patrol service may be well justified. Fire departments will usually dispatch immediately on any alarm call.

Digital communicators are only as reliable as the telephone lines they are connected to. As part of many security installations, telephone line connections to the house are encased in armored cable. A telephone line monitor used in the security system will sound a local alarm if the phone lines are tampered with, cut, or disconnected. Long-range radio is another popular option. Some telephone companies offer supervised connections that will automatically generate a trouble signal if the lines are interrupted.

Notes

Notes

Notes

CHAPTER 9

Lighting

Lighting plays a critical role in the security, comfort, convenience, and safety of a household. With home automation controlling the lighting, numerous lighting schemes can be managed for any number of events. For example, upon arriving home, the homeowner can unlock the front door with a remote. Opening the door may in turn prompt the home automation controller to turn on selected lights so that homeowners do not enter a dark house. If someone approaches at night, all security lights can switch on or perhaps only those in the area where movement is detected.

In new construction as well as retrofitting existing homes, lighting controls are extremely important. Since most light modules and controllers use power lines to carry signals, there is no need for extensive rewiring, breached walls, or unsightly fittings. Automated control switches are designed to fit into standard wall boxes and are easy to install.

Lighting System

The common concept of a lighting system consists of components such as lamps, ballast, and luminaires (a complete lighting unit). While this definition may be adequate for some purposes, it fails to consider other critical elements. Environmental, human, and task components are also part of the lighting system in a broader, more comprehensive sense. Since a lighting system is actually a complicated interaction between the lighting hardware, the environment, human vision, and the task, each element needs to be understood in detail.

Lighting Components

THE POWER SOURCE

The power source provides the source of energy to the lighting system. Variations in the power source's voltage can reduce lighting levels or, in some cases, cause problems with ballast, both in starting and in operation.

POWER CONTROLLER

The power controller modifies or regulates the operation of the lighting system. Controls can affect both the amount of time that the lighting system operates and the amount of power that it consumes while operating. Because the energy equation is Energy = Power × Time, a major component of lighting energy can be reduced through proper use of switching and dimming controls.

POWER REGULATOR (BALLAST)

The ballast provides the starting voltage and regulates the current to gaseous discharge lamps. The ballast determines the input wattage and the light output of the lamp, and is matched to the specific lamp type. The lamp-ballast combination is the primary factor affecting the overall efficiency of a lighting system.

LIGHT SOURCE (LAMP)

The lamp is the lighting system component that generates the light. Lamps use various physical phenomena to produce light, and each type has its own starting and operating characteristics. For this reason, different lamps are generally not interchangeable within the lighting system. In some cases, suitable substitute lamps can be used without changing other lighting system components.

OPTICAL CONTROL (LUMINAIRE)

The luminaire determines the way in which light is distributed. Luminaires function as the housing for the light source, and they include optical assemblies consisting of reflectors and lenses to direct the light from the source through space. Designed for specific lamp types, the optical components of luminaires are product-specific, and the use of different lamps will change the light distribution of the luminaire.

ENVIRONMENTAL COMPONENTS (ROOM FINISHES)

The reflectance and textures of room finishes affect lighting levels and the apparent brightness of the room. Dark finishes and heavy textures absorb light so that it is not reflected back into the room. Light finishes help to reflect more light into the space, thus making the lighting system more efficient. The effect of lighter wall and ceiling finishes is to increase the measured light level and to make the space appear even brighter.

SPATIAL ENVELOPE (ROOM BOUNDARIES)

Room boundaries have the greatest effect on how a space is perceived and how a lighting system will perform. Since long, narrow spaces have different proportions than do large, open areas, the efficiency of selected lighting systems will depend largely on the geometry of the room. The *room cavity ratio* (*RCR*) is the ratio of the surface area of the walls to the floor area and is often used in lighting calculations to determine how well luminaires will perform for given room proportions. In addition, spatial characteristics strongly influence how we react to a space, and the lighting components must interact with the envelope to help define and create the way in which the space is perceived.

WINDOWS AND SKYLIGHTS (FENESTRATION)

Windows and skylights admit daylight into a space. Natural lighting has an effect on the perception of time and place, which is psychologically beneficial. Windows present a challenge due to the potential for glare and/or contrast between the windows and the surrounding space. Diffuse skylights can contribute significant amounts of light into a space, offsetting the need for electric light.

Task Components

TASK FINISHES

Task finishes are the component of the lighting system that allows us to see and recognize the objects and the materials of which objects are made. Understanding the properties of the task surface is important when you are considering proper lighting solutions.

TASK (OBJECT) SIZE

The physical size of the task and the distance of the task from the observer determine the size of the visual task. Larger objects can be seen at longer distances. Small objects must be relatively close for the eye to discern them in detail.

TASK BRIGHTNESS

Exitance is the measured value of the brightness of a surface. It is equal to the amount of light falling onto the surface multiplied by the reflectance of the surface. Luminance is also a measure of the brightness of a surface but considers the direction in which light leaves the surface. A person's perception of brightness depends on the adaptive state of the eye. For example, an automobile headlight seems very bright at night, but during the day, that same headlight is barely noticeable.

CONTRAST (BRIGHTNESS RATIOS)

Surface contrast is the relationship between the luminance of an object (e.g., print) and that of its immediate background (e.g., paper). Tasks with high surface contrast are easier to see than tasks with low surface contrast. *Outline contrast* is the relationship between the luminance of the immediate background (e.g., paper) and that of the surrounding area (e.g., desk). Excessive outline contrast may make the task harder to see.

SPEED AND ACCURACY

Tasks that must be performed efficiently and accurately require fairly high luminance and contrast. Tasks for which time is not a critical factor can be performed under conditions of low luminance and low contrast. Reading a newspaper on the train is an example of a low-luminance, low-contrast task in which time is not a critical factor. Filling a prescription in a pharmacy is an example of a task requiring fast and accurate visual performance.

Because the performance of a lighting system depends on a wide range of hardware, environmental, human, and task components, lighting design is a complicated issue. Understanding these components makes it possible to design an appropriate, energy-efficient solution that meets the needs of the homeowner.

Types of Lights

Luminaires come in several types. Incandescent, fluorescent, and halogen are the most popular residential choices. Each of these lights has advantages and disadvantages. Frequently, the lighting schemes for individual rooms include two or even all three of these lighting types. Again, the key to the design depends on the needs of the homeowner(s).

Incandescent

Incandescent lights are the various bulbs we have been familiar with since the invention of electric light. Light is generated by sending electric current through a filament seated inside a glass enclosure (the bulb) until the filament glows white hot. Different shapes and sizes, as seen in Fig. 9.1, are used for a variety of needs, both indoors and outdoors. From mood lighting to task lighting, general illumination, and spot- or floodlighting, there are incandescent bulbs to fit almost any fixture and purpose.

The efficiency of incandescent bulbs is generally low. The amount of energy needed to power each light is relatively high when compared to the lumens generated. Improved incandescent lamps are available which generate higher light levels at lower levels of energy consumption. Overall, incandescent lights are the most widely used bulbs in residential buildings, but are probably the least efficient choice.

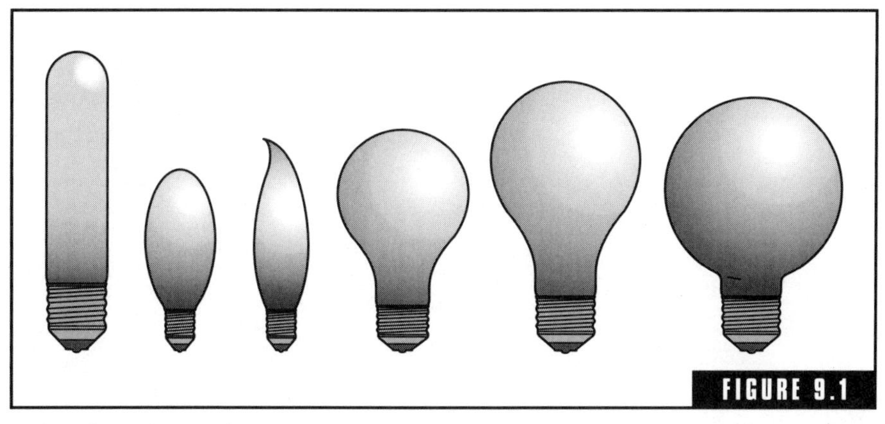

FIGURE 9.1

Various incandescent lamps.

Fluorescent

Fluorescent lamp technology has made tremendous advances over the past few years, with today's products providing greater energy efficiency, improved color rendition, and a greater selection of color temperatures. These improvements are due in large part to the use of rare-earth phosphors replacing older halophosphors used in standard "cool white" lamps. To a lesser degree, efficiency improvements arise from the more widespread use of smaller-diameter lamps, which can also increase luminaire efficiency and improve light distribution patterns.

Rare-earth phosphor compounds are selected for their ability to produce visible light at the most sensitive wavelengths of the red, blue, and green sensors of the eye. When compared with conventional halophosphors, such as cool white, rare-earth phosphors produce better color rendering and higher efficacy, while improving lumen maintenance characteristics. For reasons of lumen maintenance, rare-earth materials are required in small-diameter lamps, such as the compact fluorescent types (Fig. 9.2). Rare-earth phosphors also raise lumen output by up to 8 percent above that of conventional halophosphors. Rare-earth phosphor lamps are available for most fluorescent lamp configurations and come in a wide range of color temperatures.

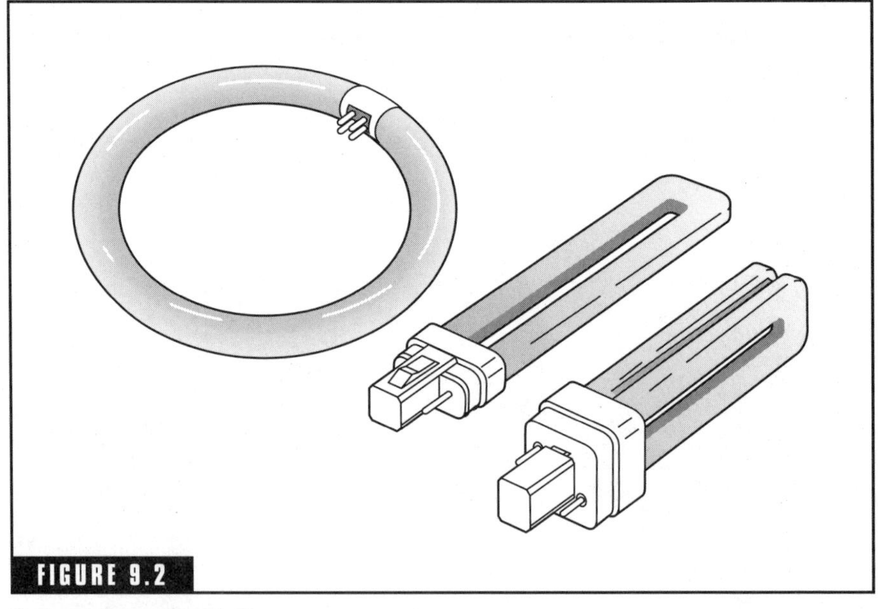

FIGURE 9.2

Compact fluorescent bulbs.

A fluorescent lamp is a glass tube, coated on the inside surface with a phosphor material and filled with argon or an argon-krypton gas mixture. The lamp also contains a small amount of mercury. When a suitable high voltage is applied across the electrodes located at each end of the sealed tube, an electric arc discharge is initiated, and the resulting current ionizes the mercury vapor. The ionized mercury emits ultraviolet (UV) radiation, which strikes and excites the phosphor coating, causing it to glow, or fluoresce, and produce visible light. About 22 percent of the energy used by the lamp is converted to light. The exact makeup of the phosphor coating determines the color properties of the light output of the lamp.

Fluorescent lamps require a ballast to initiate the discharge and regulate the electric current through the lamp. For optimum performance, the particular ballast must match the current requirements of a specific lamp. Dimming characteristics vary depending on the lamp type. For instance, the dimming of compact fluorescent lamps is currently difficult and expensive, although new dimmable products are being developed. Near full-range dimming is possible with 4-ft-long T8, T10, and T12 lamps (Fig. 9.3) with special ballast and dimming equipment. The low end of the dimming range is from 1 to

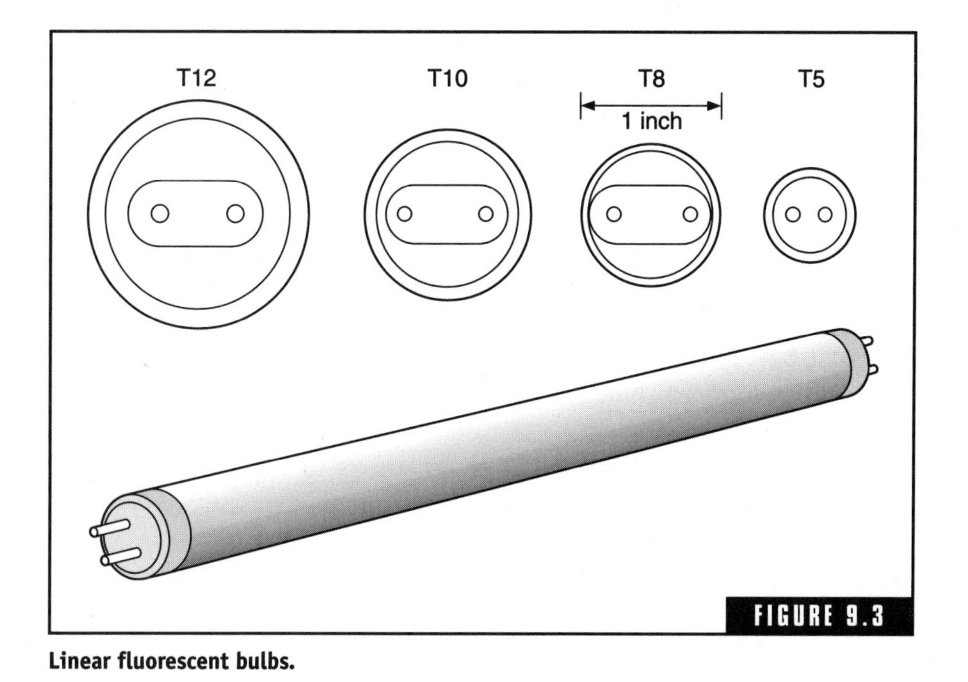

Linear fluorescent bulbs.

25 percent of full light output, depending on the lamp type and dimming ballast.

ENERGY-SAVING (ES) LAMPS

In response to the energy crisis of the 1970s, lamp companies introduced "energy-saving" lamps with krypton added to the gas fill. These lamps draw less power than standard lamps, usually about 35 watts (W). Because these lamps can be operated by standard ballast, they can be readily substituted in existing lighting systems. This reduces input wattage by about 12 to 15 percent, with a lumen output reduction of 18 to 20 percent for a two-lamp system. The resulting reduction in light levels is generally acceptable to most users, although the lamp color may not be as desirable.

Energy-saving lamps are more sensitive to low temperatures than standard lamps, with a minimum starting temperature of about 16°C (60°F), as opposed to 10°C (50°F) for standard 40-W lamps. Energy-saving lamps are not recommended for dimming applications.

Tungsten-Halogen

Tungsten-halogen (also called halogen) lamps use enhancements to conventional incandescent technology to provide improved energy efficiency. They provide a whiter, brighter light and longer lamp life than do conventional incandescent products. Tungsten-halogen lamps are also more compact, providing better optical control and enabling smaller reflector and luminaire designs.

Although tungsten-halogen lamps cannot compete with fluorescent lamps in terms of energy efficiency or lamp life, they are an appropriate choice for situations requiring directional or controlled light. Halogen lights are often used as accent or display lighting, or where full-range dimming is required. There are two categories of tungsten-halogen lamps: line voltage tungsten-halogen products and low-voltage tungsten-halogen products.

Incandescent lamps produced light by passing electricity through a tungsten filament, heating it so that it glows or incandesces. Since light generation is a by-product of heat, incandescent lamps are inefficient light producers, with only about 10 percent of the energy consumed being turned into light. In spite of this fact, incandescent lamps are effective for certain applications. Since the filament is very small and is, in effect, a point source, the light itself can be directed easily.

Tungsten-halogen lamps use a halogen gas, usually iodine or bromine, to suppress filament evaporation by a chemical regeneration process known as the *halogen cycle*. During lamp operation, the halogen gas combines with tungsten molecules that have evaporated off the filament. The evaporated tungsten molecules are redeposited onto the filament, instead of onto the bulb wall. Consequently, lamp lumen depreciation due to bulb wall darkening is practically nonexistent. Depreciation occurs, due to filament degradation, but is significantly lower than that in other incandescent lamps.

The halogen regenerative process requires operating tungsten-halogen lamps at extremely high temperatures, increasing lamp efficacy slightly, while producing a brighter, higher-color-temperature light. To withstand these high temperatures, tungsten-halogen lamps have special glass envelopes—usually quartz (see Fig. 9.4). Lamps with quartz bulb walls require special handling, as quartz materials are extremely sensitive to oil and dirt from human skin, which can cause bulb wall deterioration and premature lamp failure.

Reducing the voltage flowing to halogen lamps produces a decrease in lumen output and an increase in lamp life. As with all incandescent lamps, simply lowering the voltage across the lamp can dim tungsten-halogen lamps. Full-range dimming is relatively easy and inexpensive. The color temperature varies over the dimming range, becoming warmer as lamps are dimmed. Generally, operating halogen lamps in a dimmed mode extends lamp life, although the increased lamp life does not follow standard incandescent lamp curves.

INFRARED (IR) REFLECTING FILM LAMPS

Up to 90 percent of the energy radiated by incandescent lamps, including tungsten-halogen lamps, is in the form of invisible infrared heat. However, some of this infrared energy can be indirectly converted to light through the application of a diachroic film coating to the tungsten-halogen lamp (or capsule, in the case of PAR lamps). This coating consists of several layers of a microthin optical material. Heat energy is reflected back to the lamp filament while visible light passes through the bulb wall. The reflected infrared, in turn, further heats the tungsten filament. Consequently, the necessary operating temperature for the halogen cycle is maintained with less power.

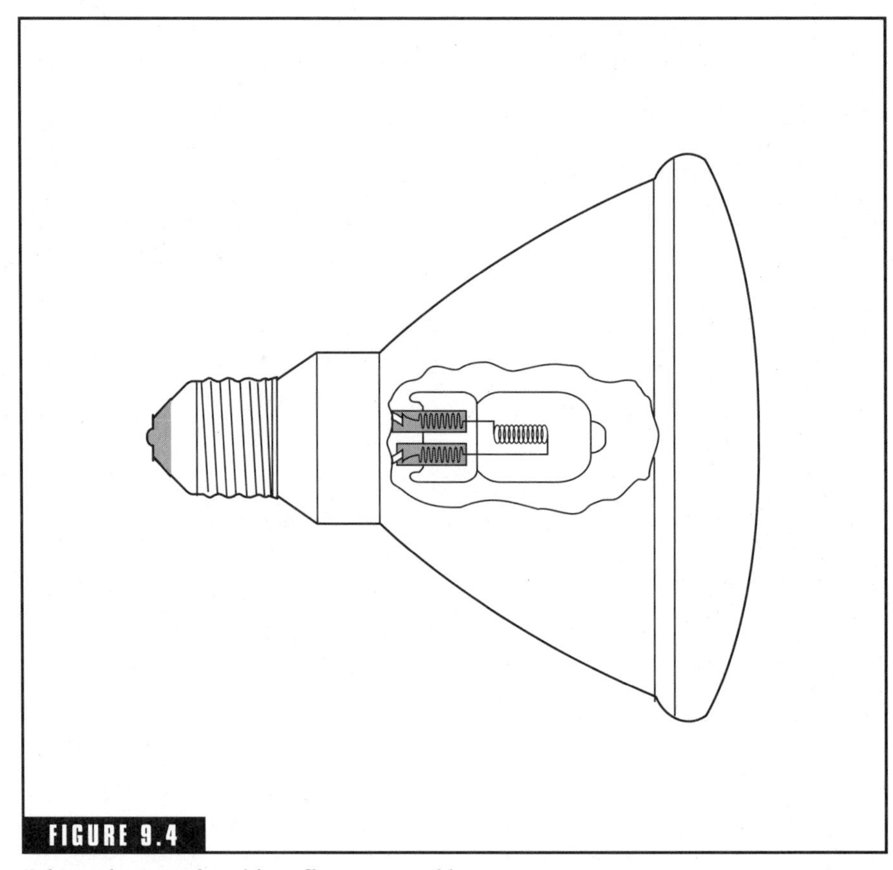

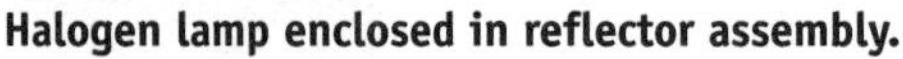

FIGURE 9.4

Halogen lamp enclosed in reflector assembly.

Infrared reflecting lamps (Fig. 9.5) provide all the benefits of standard tungsten-halogen lamps, including low lumen depreciation and high-quality light. The energy-saving potential with the double-ended and PAR lamp configurations is spectacular. In some cases, replacing a standard incandescent lamp with a halogen lamp can reduce the wattage by 30 to 50 percent with no visible difference in light output or quality.

Since halogen lamps are relatively low-efficiency, their use should be restricted to applications requiring their unique characteristics. A common misapplication of halogen lamps is seen in the general lighting of large spaces. Ideally, halogen should be used only in applications where high footcandle levels are needed in small areas. Otherwise, light schemes should take advantage of other, more energy-efficient options. Illuminating larger areas is more the province of one of the following:

- Compact metal halide and white high-pressure sodium lamps in general lighting, wall washing, and display lighting situations
- Compact fluorescent lamps, especially in area lighting and wall-washing situations
- High-wattage, high-intensity discharge (HID) lamps for a high-lumen, high-efficacy source

Low-Voltage Tungsten-Halogen

Originally developed for automotive and aircraft applications, low-voltage tungsten-halogen lamps normally have extremely compact lamp envelopes and filaments. Low-voltage halogen lamps provide excellent optical control, creating intense, focused beams. In addition, these lamps offer traditional halogen color temperature and lamp life advantages.

Although these products are more efficient than equivalent line voltage lamps, they are still less effective than other advanced lighting technologies. Until recently, low-voltage operation required a separate transformer to provide the proper power supply. Many low-voltage halogen products are now supplied with internal transformers to make installation easier.

Most low-voltage halogen lamps use a compact quartz-glass envelope or *bud* with two vertical pin terminals. As is the case with line voltage halogen lamps, it is essential not to touch the bud with human

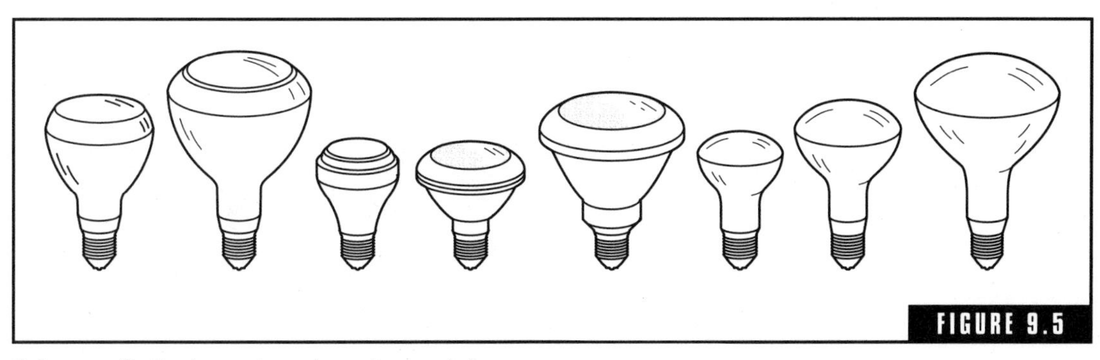

FIGURE 9.5

Halogen reflector lamps in various sizes and shapes.

skin. The oil deposited on the bud by human skin causes hot spots and drastically reduces the life of the unit.

APPLICATIONS

The most common applications for low-voltage halogen lamps are as projector or reflector lamps. In these lamps, the halogen bud has a reflector of aluminum or glass. There are three distinct types of projector lamps: multimirror reflector (MR) lamps, aluminum reflector (AR) lamps, and low-voltage PAR-36 lamps. Figure 9.6 shows several types of low-voltage halogen lamps.

Multimirror reflector lamps have dichroic-coated faceted-glass reflectors, and they are available in many wattages and beam spreads for a variety of accent lighting applications. The most popular low-voltage lamps are MR-16 bipin lamps. Originally developed as slide projector lamps, the architectural versions are 20- to 75-W, 12-V lamps. The reflector is made of faceted glass, coated with a dichroic film that reflects visible light and transmits infrared energy through the back of the lamp, making the MR-16 beam an inherently cool-beam lamp.

The various MR-16 lamps perform as very narrow spots, narrow spots, narrow floods, and wide flood beam spreads, with beam spreads determined by the size and orientation of the facets on the reflector face of the lamp.

PAR-36 LAMPS

Low-voltage PAR-36 lamps have been popular with lighting designers for many years. Today's halogen buds enclosed within PAR-36 glass envelopes provide similar performance to that of standard incandescent PAR-36 lamps, but with improved color rendering, longer lamp

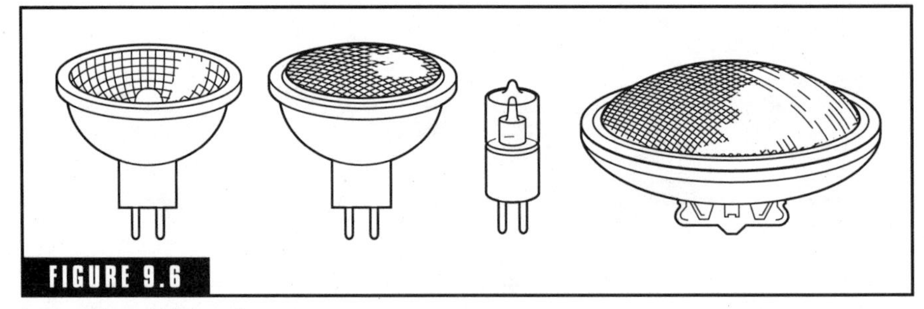

FIGURE 9.6

Low-voltage halogen lamps.

life, and improved energy efficiency. In addition, low-voltage PAR-36 lamps allow exceptional beam spread control.

Fixtures

Virtually all the lighting types mentioned in this chapter are adaptable to the many different kinds of fixtures that can be used in and around the house. Depending on the effect desired for a particular room or section of a room, specific fixtures are used to fulfill those desires. Some of the more popular fixtures are discussed in this section.

Recessed Lighting

Recessed fixtures are an excellent source of general lighting as well as task and accent lighting. Fixtures are hidden up in the ceiling so that these fixtures provide light without attracting any attention.

Design experts recommend using downlights for general lighting needs. Downlights in groups of four or more, evenly spaced 6 to 8 ft apart, provide wide-coverage ambient lighting. Generally, one fixture for every 25 ft^2 will be sufficient. Placing downlights above the front edge of a countertop or other work areas will illuminate the workspace directly in front of the homeowner. Downlights installed every 8 to 10 ft in the exterior eaves of a home can create a dramatic effect by washing exterior walls with a soft light.

Using recessed lighting to accent artwork and architectural features can create dramatic effects. Placing eyeball fixtures 6 to 8 in from a wall and 12 to 30 in apart can emphasize brick and stone walls. A recessed shower light over tubs and showers enhances safety during bathing and showering.

Recessed downlights are a popular choice for general lighting in kitchens because of their relatively unobtrusive appearance. Recessed fixtures, by their very nature, create a large hole in the ceiling and allow a significant amount of air to pass from the room into the space above the ceiling. This frequently creates problems with insulated ceilings. Heated air from the kitchen is lost in the air escaping through the fixtures. Along with heat, water vapor generated in large amounts in kitchens travels through the recessed fixture directly into the attic or wall cavity, where moisture can condense on cool surfaces, such as roof sheathing. This could lead to mold, mildew, and structural decay.

Cathedral ceilings are especially susceptible to this moisture problem because of their limited ventilation space.

If recessed lighting is to be installed in high-moisture areas (kitchens and baths), specify airtight models that have been tested for low air leakage according to ASTM 283E. The results should show no more than 2 cubic feet per minute (ft^3/min, or cfm) of airflow at 50 pascals (Pa) of pressure. Also, consider surface-mounted fixtures, such as track lights, instead of recessed ones.

Track Lighting

Track lighting refers to a flexible system of lighting composed of several elements. Normally, track lighting utilizes a length of electrically fed linear track, an electrical feedbox, and one or more heads, each containing a lamp. Directly mounted to or recessed into the ceiling, the track functions as a base for the lighting heads.

Lighting heads are movable and can direct light in precisely controlled patterns. In some lighting schemes, track lighting is suspended from ceilings while in others it is mounted on walls for indirect lighting. Monopoint mounts allow a single-track head to be mounted on an outlet box instead of a track.

Track is available in lengths of 2, 4, or 8 ft. Lighting heads come in many shapes, sizes, and styles (Fig. 9.7). Track lighting is available to support the use of standard incandescent, halogen, and low-voltage halogen lights.

Wall Sconce

Wall sconces used in various locations can service different needs for general lighting, task lighting, and accent lighting. Sconces placed every 8 to 10 ft horizontally and at a height of 5$^1/_2$ to 2 ft from the floor along a hallway should provide adequate lighting.

Using sconces on either side of an entryway or fireplace provides accent lighting to the area. In great rooms featuring high, vaulted ceilings, high-output wall sconces can provide the main source of general lighting. Sconces can be used in bedrooms to replace table lamps on either side of the bed. These installations provide task lighting for reading and can be dimmed for mood lighting as well. Sconces used in smaller rooms, such as bathrooms (Fig. 9.8), provide general lighting; and dimmers permit adapting the light to specific tasks, such as shaving or putting on makeup.

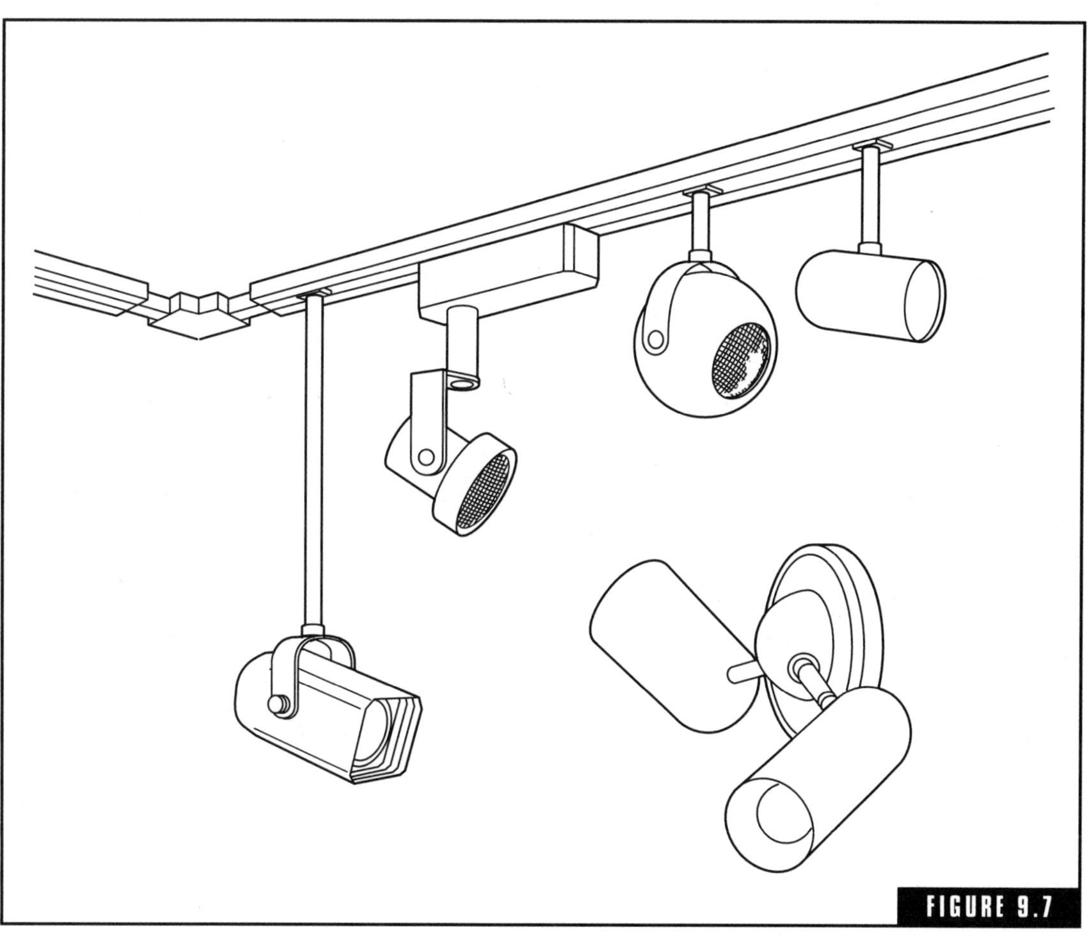

Example of track lighting.

Remote-Source Lighting

Remote-source lighting refers to two types of new lighting technologies: fiber optics and prism light guides. The use of remote-source lighting is steadily increasing, and these technologies may eventually replace conventional lighting in several indoor and outdoor applications.

Nontraditional lighting applications such as in data communications, medical imaging, specialized lighting for dentistry, microscopes, cameras, and instrument displays already employ fiber-optic systems. Prism light guides, first manufactured in 1983, are relatively new and not yet as commonly known as fiber optics. Fiber optics and prism light guides are gaining acceptance because of their unique features and benefits.

FIGURE 9.8

Sconces provide general and task lighting in bathrooms.

The ability to light multiple locations with one source leads to several benefits. The light emitted from remote-source lighting systems is often called *cold light* because it contains no infrared energy. The system also filters out the ultraviolet energy emitted by the source. Objects sensitive to infrared and/or ultraviolet light, such as paintings, textiles, or antiques, can be safely lighted by using remote-source lighting systems. There is no heat generated to discolor room surfaces or increase room temperature.

Because remote-source lighting systems have no electrical parts, they can be safely used in hazardous environments. Workshops, garages, or storage areas containing flammable gases are afforded additional safety when lighted with these systems. Wet locations such as pools and spas lighted with remote-source systems eliminate the electrical hazards inherent with traditional lighting. Also, the light guides carry light instead of electricity, so they cannot conduct electromagnetic interference (EMI). Thus, by remotely locating the electronics (lamp and ballast), remote-source lighting systems are ideal in areas with EMI-sensitive equipment.

MAINTENANCE

Replacing traditional lighting with a multiple-outlet remote-source system means that there's only one lamp to replace. The light source can be put in a convenient, safe, accessible place. High-ceiling applications, such as downlighting recessed in a cathedral ceiling, can be easily maintained by locating the lamp and electronics in a convenient closet.

ENERGY CONSUMPTION

Remote-source lighting installations with a high-efficacy source (such as HID lamps) are more efficient than an installation using multiple incandescent lamps. Less demand is placed on the air conditioning system since remote-source lighting systems emit less heat than conventional systems. For example, replacing halogen track lighting with a fiber-optic system can result in energy savings of as much as 50 to 70 percent. Such savings have been typical of remote-source lighting systems.

Working with Fiber-Optic Cable

All glass and some plastic fiber systems are preassembled by the manufacturer according to job specifications. No cutting or polishing of the fiber bundles is needed on site. However, the use of solid-core plastic fibers is increasing, resulting in more on-site assemblies. Optical fiber (Fig. 9.9) is composed of a light-transmitting core inside of a cladding that traps the light within the core, causing total internal reflection. Fiber has two basic types: multimode and single-mode. Multimode fiber means that light can travel over many different paths (called *modes*) through the core of the fiber, which enter and leave the fiber at various angles. Two types of multimode fiber exist, distinguished by

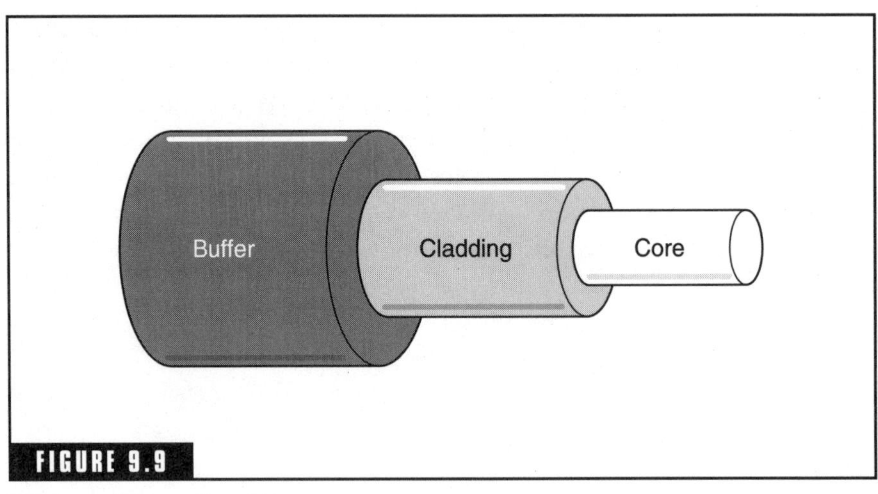

FIGURE 9.9

Solid-core fiber-optic cable.

the index profile of their cores and how light travels in them. Single-mode fiber shrinks the core size to a dimension about 6 times that of the wavelength of the fiber. The result is all the light traveling in only one mode. Consequently, modal dispersion disappears, and the bandwidth of the fiber increases to at least 100 times that of graded index fiber.

The use of plastic fibers for installation increases design flexibility and lowers cost. The large solid-core plastic fiber is supplied on a spool. It is cut to length as needed, and both ends are buffed or polished for light transmission. The success of glass or plastic fiber optics depends on the lighting design's requirements and the installer's competency.

Exterior Lighting

Exterior lighting not only enhances the beauty of the home, but also provides safety and security for the homeowners, guests, and property. Well-designed exterior lighting systems make dramatic aesthetic statements while illuminating house numbers, locks, and changes in surfaces and levels. All the various types of lighting (including fiber-optic) are adaptable to exterior use. Combinations of general lighting, accent lighting, and task lighting will provide for the homeowner's needs around the outside of the house and throughout the yard.

General Installation Tips

When you are planning exterior lighting for a home, there are some rules of thumb that will guide the installation:

- Size fixtures in proportion to the size of the home. Remember, a wall fixture 2 ft tall does not necessarily look out of place on a three-story home, especially when viewed from 30 ft away (at the street). Selecting a small fixture for a larger home will take away from the presence of your home.
- Position outdoor lights to ensure that unwanted light does not shine toward neighboring properties. Aim floodlights at a 45° down angle.
- Always caulk around fixtures that are attached to the exterior of the home. This step avoids problems created by moisture settling between the fixture and the wall.
- To dramatically accent landscaping and enhance the appearance of the home, illuminate pathways and steps with spreads of light.
- From a security standpoint, it is critical to light all sides of the house and floodlight shrubbery and building offsets that could shelter burglars.

Lighting Control

There are a variety of means for automatically controlling lights and appliances. All major home automation protocols enable control of lights and electric appliances anywhere on the property. Most are compatible for use with or without a computer. For example, each device connected to an X-10 system has its own receiver, which can be an X-10 light switch replacing a standard wall switch, or it might be a small device that plugs into a standard outlet.

Control of such devices can be accomplished from a PC or Mac, or a simple timer can be used for control. One of several remote devices, including a wireless controller, can be employed. These controllers have been used for years to turn lights on in the early evening, dim them, and turn them off at a preset time. Outdoor lights can also be controlled by self-contained light sensors, which turn on at dusk and off at dawn.

Also X-10 devices can be used as part of a home security system. X-10 USA and other companies make a variety of motion detectors and other sensors that turn on lights, set off alarms, or call the police. The X-10 protocol does not require any special wiring, sending signals through the home's regular electric wires. Optional wireless transmitters emit radio frequencies or infrared signals that trigger a controller plugged into the wall. Other methods are used to transmit data throughout the home. CEBus (Consumer Electronics Bus) supports a variety of transmission methods—including power line carrier, coaxial cable, twisted-pair wiring, and infrared—to transmit data to and from appliances.

Computer Controls

Lighting systems for both interior and exterior uses can easily be tied into the central home automation system. Control systems can be broadly divided into two groups depending on whether the system intelligence is centralized or distributed. Most of, if not all, home control systems currently available on the market fall into the first category. System intelligence is centralized either in a personal computer or in a separate microcontroller (which may or may not be interfaced to a personal computer).

The primary interface is normally the personal computer, a keypad, a "soft key" LCD panel, a touch screen, a handheld remote, a telephone interface, a television, or even a voice command. The commands from the central controller are transmitted to various local systems, such as lights, using dedicated wiring, PLC house wiring, wireless signals (RF or IR), or a combination of these media.

Distributed home control systems are in various development stages within the home automation industry. Two communications protocols (CEBus and LonWorks) hold promise for the future of home automation.

Computer controllers that plug into serial ports of PCs and Macs turn lights on and off or dim lights based on whatever is preprogrammed into the computer. The more advanced automation systems not only can take preprogrammed instructions, but also can accept remote commands. Of course, all computer controllers are used in conjunction with software. Some software is sold with automation controllers, and other software is independent of branded controllers and can be used with most products. A single command sent from any device back to the controller could trigger an entire series of events. For example, a motion sensor controlling perimeter lighting in the yard can also cause all the lights on the second floor of the house to go on, giving the appearance of human response.

Sensors

OCCUPANCY SENSORS

Occupancy sensors react to variables such as heat and/or motion by turning lights on or off. The sensors turn lights on when the presence of people is detected; then, after an adjustable predetermined period during which people are not detected, they turn the lights off. There are a variety of sensor types and placement strategies.

PASSIVE INFRARED (PIR) SENSORS

Passive IR sensors are the most commonly used sensors both inside and outside the home. These sensors are sensitive to the infrared heat energy emitted by people or cars. Triggering occurs when they detect a change in infrared levels, such as when a warm object moves in or out of view of one of the sensors. PIR sensors are passive: They detect radiation but do not emit it. They are maximally sensitive to objects that emit heat energy at the wavelength emitted by humans. Strictly line-of-sight devices, they cannot "see" around corners, and a person will not be detected if there is an obstruction such as a partition, tree, or bush between the person and the detector. PIR sensors are fairly resistant to false triggering.

ULTRASONIC SENSOR (US)

Ultrasonic sensors emit high-frequency sound and listen for a change in frequency of the reflected sound. They cover larger areas than PIR sensors and are noticeably more sensitive, but are also more prone to false triggering than PIR sensors. Air motion from a person passing an open door or the on-off cycling of an HVAC system may trigger a poorly located or adjusted sensor. Consequently, ultrasonic sensors are not usually recommended for outdoor use. Several manufacturers offer hybrid sensors. These sensors use both infrared and ultrasonic technologies simultaneously. They combine the sensitivity of ultrasonics with the passive infrared unit's resistance to false triggering.

CEILING-MOUNTED SENSOR

Ceiling-mounted units employ an independent controller and/or power supply. They may be mounted high on a wall or in a corner as well as on the ceiling.

WALL-MOUNTED SENSORS

Wallbox sensors are primarily designed to replace common wall switches. Both types of sensors are available with either PIR or ultrasonic sensing units. They can be combined to cover an odd-shaped or large room.

TIME-BASED CONTROLS

Elapsed-time switches typically fit into or over a standard wall switch box and allow occupants to turn lights on for a period set by either the occupant or the installer. Lights go off at the end of the period, unless the occupant has restarted the time cycle.

CLOCK SWITCHES

Clock switches turn lights on and off at preset times, regardless of occupancy. They can be mechanical or electronic. There are several varieties, such as 24-hour and 7-day time switches. Some clock switches have a feature that automatically compensates for changing ambient light.

PHOTOSENSORS

Photosensors respond to changes in light. They turn lights on when ambient light falls below a preset level and off when that level is exceeded. Like time switches, photosensors have been used for many years to control outdoor lighting. They can respond to overcast conditions during the day and provide safety and security when more lighting is needed unexpectedly. Some have time-delay devices to help prevent rapid off-on cycling on partly cloudy days.

As is the case with so many components of home automation, the selection of lighting elements and controllers largely depends on the needs and lifestyle of the homeowners. Thoroughly discussing lighting alternatives with the homeowners prior to construction will ensure lighting schemes that provide an environment which conforms to the homeowner's changing needs.

Notes

Notes

More Uses for Home Automation

As homeowners become more aware of home automation and the benefits that can be realized through these systems, the demand for products increases. The number of manufacturers is increasing, and each is developing new devices to work in conjunction with home automation. This chapter deals with some of the major ancillary devices, which are compatible with home automation systems.

Outdoor Watering Systems

Residential watering systems have been around for decades. From using a hose to water plants, shrubs, and lawns to in-ground sprinklers, homeowners have constantly searched for easier ways to take care of watering tasks. Home automation, by a strict interpretation of its definition, refers to a system or subsystem that functions without the participation of the homeowner. This interpretation includes thought.

True automation requires the system or subsystem to operate without the homeowner's pushing a button, flipping a switch, or even thinking about the operation of the system. Today, this type of automation is possible, even for tasks such as watering the yard.

Installing a Full Yard Watering System

Before you connect the sprinkler (or irrigation) system to the automation system, it is important to understand how the irrigation system is installed and how it works. As with any other stage of construction, the first requirement of installing an irrigation system is to prepare a takeoff of the materials. To promote uniformity, reduce errors, and expedite the work, implement a standard system for doing takeoffs. Incorporate a system of checks wherever possible, to prevent omissions and errors; it is cheap insurance when compared to the cost of furnishing a dozen sprinkler heads not accounted for in the bid. A checklist of items normally found in irrigation systems should always be included as part of the takeoff. Some contractors use a standardized takeoff form that incorporates a checklist. Table 10.1 is an example of common items found on an irrigation checklist.

Riser Types

There are many ways to install a sprinkler head. One of the most common ways for residential use is to simply install the sprinkler on a short length of threaded pipe. Although inexpensive, this riser type does have some disadvantages. If the nipple installed under the sprinkler is rigid PVC plastic or metal, the riser normally will not break. However, under stress either the sprinkler head or the fitting under the nipple will suffer, resulting in breakage. Both of these components are more expensive and more difficult to replace than the nipple.

An inexpensive solution is to use a polyethylene cutoff nipple for sprinkler risers. A cutoff nipple is a short pipe section (typically 6 in long) with multiple sets of threads on it. Simply cut it to the desired length. Because the poly material is very soft, the nipple will bend under stress and will break before either the sprinkler or the fitting.

SWING JOINT OR SWING RISER

A swing joint allows the sprinkler head location to be easily adjusted while also deflecting in order to prevent breakage. Since the sprinkler head doesn't need to be directly over the lateral pipe fitting, pipe placement is not as critical. Another benefit is that trenching and pipe installations are done more easily and quickly. Figure 10.1 shows an example of a flexible-arm swing joint. This version of a swing joint,

TABLE 10.1 Takeoff Checklist for Drip and Sprinkler Irrigation Systems

1.	Sprinkler heads—nozzles, rubber covers, internal check valves, stainless sleeves
2.	Bubblers and drip emitters
3.	Emitter distribution tube (spaghetti tube)
4.	Microsprinklers and risers
5.	Isolation valves—gate ball butterfly
6.	Master valve
7.	Control valves—latching solenoids reclaimed water
8.	Quick coupler valves—rubber-covered locking lid reclaimed water
9.	Hose bibs and keys
10.	Water meter
11.	Backflow preventers
12.	Controllers—pedestals, enclosures, junction boxes, and batteries
13.	Filters—special mesh-screen flush outlet
14.	Pressure regulators—custom range of pounds per square inch
15.	Pump station—electrical panel custom-built
16.	Air vent
17.	Vacuum release valve
18.	Pressure release valve
19.	Check valves
20.	Pressure gauge
21.	Rain switch
22.	Moisture sensor
23.	Fertilizer injector
24.	Risers for sprinklers, valves, backflow preventers, quick coupler valves
25.	Risers for isolation valves, vacuum release valve, pressure release valve

TABLE 10.1 Takeoff Checklist for Drip and Sprinkler Irrigation Systems (*Continued*)

26.	Risers for pump station, air vents, master valve
27.	Rebar and clamps for staking sprinklers
28.	Valve boxes for automatic valves, isolation valves, quick coupler valves
29.	Valve boxes for filters, water meter, pressure regulator
30.	Valve boxes for air vent, vacuum release, pressure release
31.	Valve marker
32.	Valve and wire ID labels
33.	Operation keys and wrenches for isolation valves
34.	Quick coupler valve couplers, lock-top keys, hose swivels, hose
35.	Repair tool kit
36.	Mainline pipe
37.	Mainline fittings
38.	Lateral pipe
39.	Lateral fittings
40.	Pipe couplings
41.	PVC cement, primer, and applicators
42.	Gasket lube for gasketed pipe
43.	Pipe dope, Teflon tape
44.	IPS flexible PVC hose (IPS = iron pipe size)
45.	Drip tube
46.	Drip tube fittings
47.	Clamps for insert fittings (polyethylene pipe systems)
48.	Pipe sleeves and wire sleeves
49.	Flanges, flange bolt, and gasket kits
50.	Control wire
51.	Wire splice sealers, electrical tape

TABLE 10.1 *(Continued)*

52.	Electric pull boxes
53.	Primary electrical (115- and 230-V) materials for power supply to controllers, pumps
54.	Control tubing (unusual—for hydraulic powered valves only)
55.	Drainage pipe and fittings
56.	Staking flags, caution tape, marking paint
57.	Paint
58.	Drinking fountains
59.	Corrosion protection tape for wrapping underground steel pipe

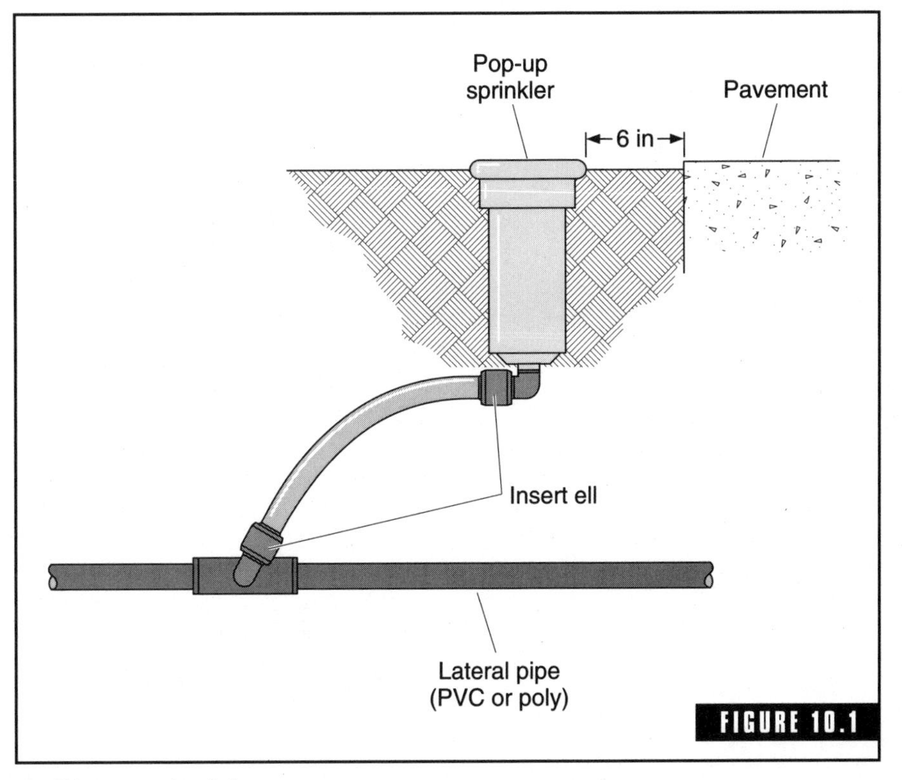

Flexible-arm swing joint.

although less expensive and easier to install, is not as durable as the rigid-arm swing joint shown in Fig. 10.2.

The rigid-arm swing joint (also known as a quadruple swing joint) is the standard riser for large sprinkler heads. This application is normally reserved for large acreage. Residential use generally dictates the flexible-arm swing joint.

ANTISIPHON VALVES

Antisiphon valves are the primary type of valves used in residential irrigation systems. An antisiphon valve (Fig. 10.3) must be installed higher than the level of the sprinkler heads or emitters that it controls. If it is not installed above the level of the heads, the built-in backflow preventer will fail. In many areas, local code requires the use of metal pipe on the inlet side of the antisiphon valve. This is because the Uniform Plumbing Code only allows PVC plastic pipe to

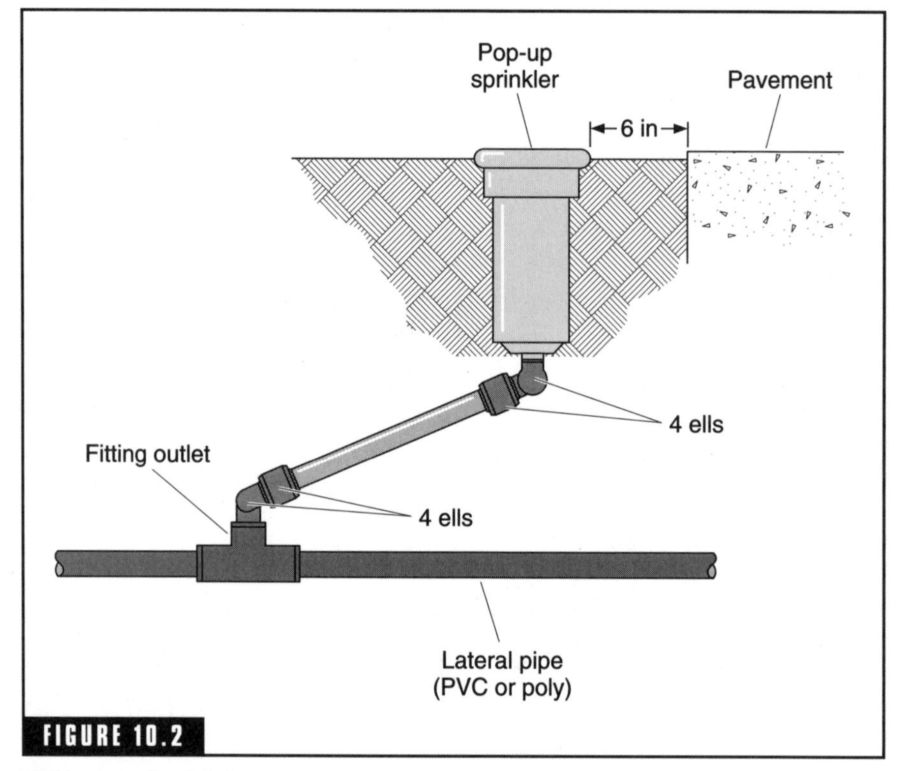

FIGURE 10.2

Rigid-arm swing joint.

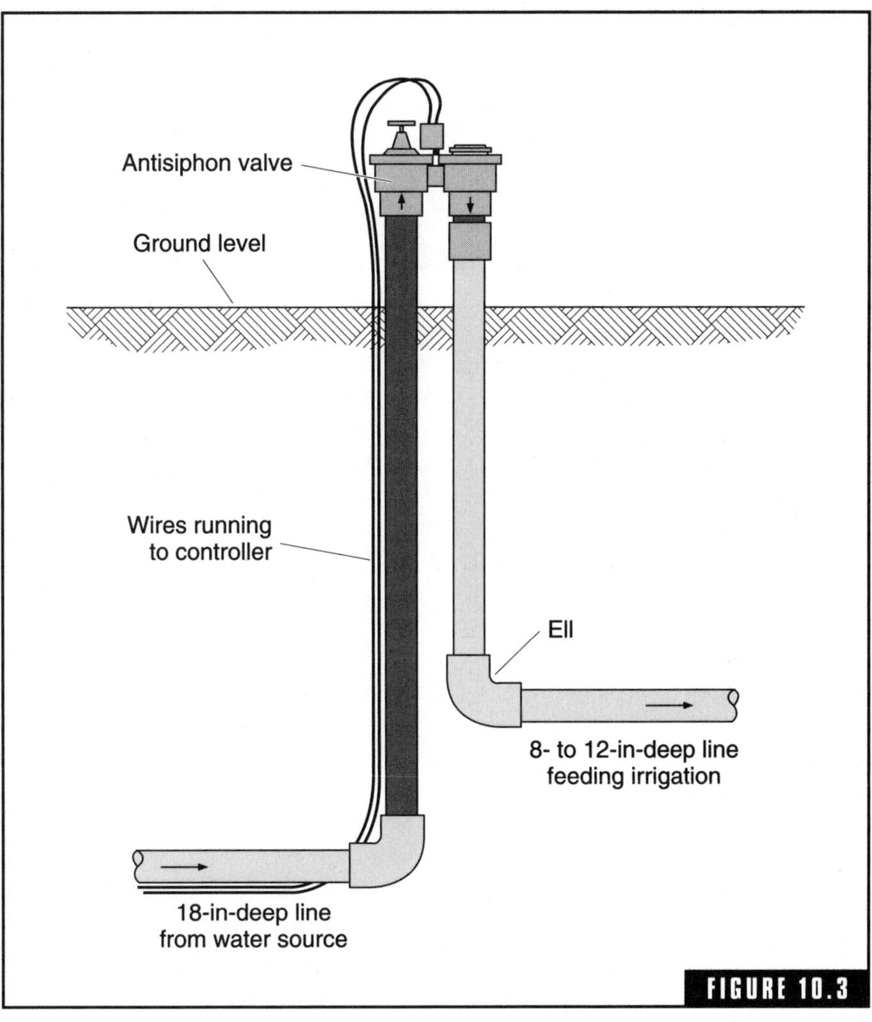

An antisiphon valve.

be used on pressure lines at least 18 in deep; anything less than 18 in deep must be metal.

Keep in mind that PVC pipe should not be left exposed to sunlight for long periods (more than a few months). The PVC breaks down when exposed to sunlight, causing structural failure. When you install the antisiphon valves on the house supply pipe (Fig. 10.4), install them only where the water supply enters the house, instead of using a separate sprinkler connection. Run a new supply pipe around the house for the valves on the opposite side, if necessary.

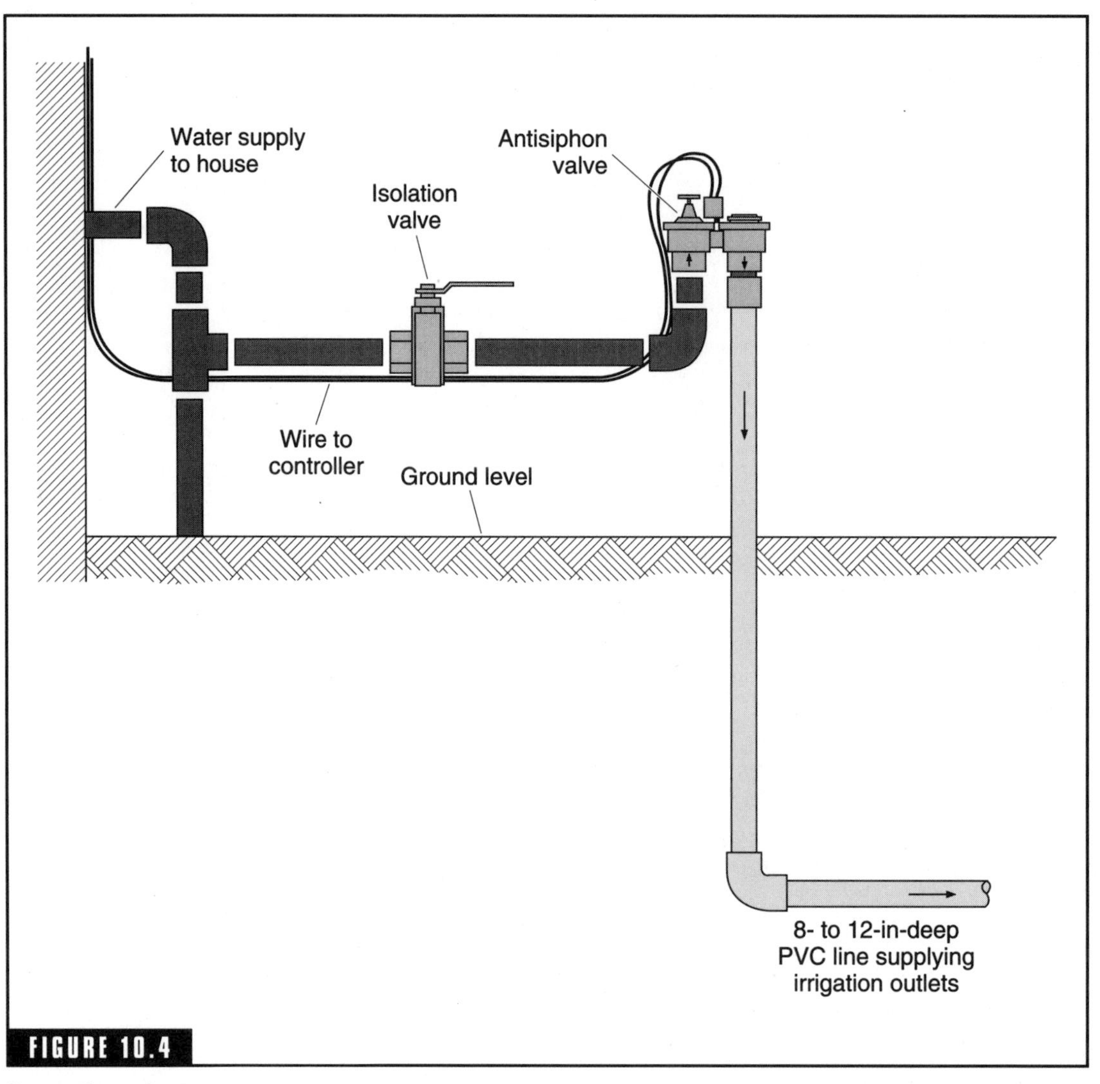

FIGURE 10.4

Connecting valve to water entrance.

Connecting the valves in this manner will eliminate problems that create restrictions in the pipe which runs through the house severely reducing water flow. Another reason for this type of installation is customer satisfaction. Sprinklers running at night will not result in water screaming through the house pipes.

By cutting into the supply pipe and installing a compression tee, you can easily tap into the water supply line leading to the house.

Remember to install an emergency isolation valve as close as possible to the tap point.

Note: Inexpensive gate valves may fail after a number of uses. Ball valves are usually more reliable and durable for this type of installation.

The size of the compression tee will depend on the size of the pipe being tapping into as well as the material the pipe is made out of. A 2-in copper pipe requires a different size of compression tee than does a 2-in PVC pipe. After you shut off the water in the supply line, cut a small section out of the pipe to be tapped. The cutout should be about one-half the length of the compression tee.

Loosen the rings on each end of the compression tee used to tighten it onto the pipe. Warp the pipe slightly, and slide the compression tee onto it. Now slide the compression tee over the other end of the pipe. Tighten the rings down, to seal the compression tee onto the pipe to complete the tap. Install new pipe going to the irrigation system into the side outlet of the tee.

BACKFLOW PREVENTER

A backflow preventer on the irrigation system is essential. In most locations this is stipulated by code to prevent toxic chemicals (fertilizers, pesticides, etc.) and animal waste from finding their way through irrigation pipes and into drinking water. Antisiphon valves normally include a built-in backflow preventer, so installations using these apparatus are covered. If antisiphon valves are not used in the installation, install a backflow preventer between the points where the water supply was tapped and the location of the sprinkler system control valves.

Wire and Splices

The control wire for the valves is laid and connected while the trenches are open. Wire should be placed at least 2 in away from the pipe and located next to or under the pipe. Never place the wire above the pipe. Thus the pipe protects the wire from damage. The size of each wire is determined from a chart provided by the valve manufacturer.

Every valve model is different. Generally, in residential systems where the wire length between the controller and the valve is less than

200 ft, no. 18 wiring will work. Number 18 is the size of wire provided in most multiwire (or multiconductor) irrigation cable. Use UF-AWG wiring for direct burial in the ground. Encase the wiring in PVC pipe to protect it from burrowing rodents.

Never connect and spark the wires against each other to test them. This can damage the circuits in many irrigation control clocks. Use an ohmmeter to test the wire continuity. Completely waterproof all wire splices; just wrapping them in plastic tape is not sufficient. Twisted-pair wire is used in these installations since it can carry more voltage than single-strand wire and provides greater flexibility. However wonderful twisted-pair wire is on the inside of the home, it has one large drawback when used for exterior purposes such as irrigation. Capillary action causes water to be drawn up into the small gaps between the wires. The water eventually is drawn into the valve's solenoid, destroying it.

Poorly waterproofed wire splices cause 90 percent of electric valve failures. Several special connectors made for underground wire splices create a waterproof splice and are the recommended way to seal splices. Most of these are simply a special wire nut and/or container filled with a very thick grease gel. The splice is pushed into the grease, and the grease flows around it, sealing it.

During electrical storms, static electricity builds up in the ground, which can surge through wires and damage valves. Most controllers are designed to pass the electrical surge through to the ground without damage. Since the surge is passed into the ground, a ground wire connected to a grounding stake driven deep into the ground is required. Without a ground wire on the controller, there is no escape path for the surge to take. A separate ground wire is needed, not the so-called ground wire in the building's electric wiring circuits.

Trench Depth

Trenches vary in depth depending on a number of factors. Plastic mainline pipe must be at least 18 in, measured from the top of the pipe to the soil surface. Installing a $1^1/_2$-in mainline pipe requires a 20-in-deep trench. Lateral trenches for home irrigation systems are usually about 10 in deep. The deeper the pipe, the less likely it is to be damaged.

In regions that experience freezing temperatures, precautions must be taken to prevent water from freezing throughout the system; pipes, valves, emitters, and sprinkler heads are all vulnerable. Installing the pipe more deeply gives it greater protection against frost. Where possible, all pipe should be installed below the soil frost line. In most cases, protection against freezing is afforded by installing drain valves on the pipe and sloping the pipe down to the drain valve so that the water can drain out. A separate drain valve is needed for the section of pipe on either side of every valve.

AUTOMATIC DRAIN VALVES

Automatic drain valves work well, but there are some considerations involved with their use. Automatic drain valves open and drain all the water from the pipes every time the irrigation system is turned off. Provision must be made for the drained water (such as a pit filled with gravel) to prevent ponding problems. Because of the automatic draining each time the system is turned off, these valves waste water. Areas with expensive water rates or with restricted water use may be better served with manual drain valves. Because the pipes must refill each time the system is turned on, automatic drain valves actually cause stress on the piping, which can lead to premature failure.

Flushing the System

Prior to flushing the system, backfill the trenches to about the halfway point. Flush the system to eliminate any objects, dirt, or sand which may have settled in the supply lines. Flush one valve circuit at a time. Be sure the tops of the risers are higher than the soil level outside the trench, to prevent the flush water from ponding around them from running back into the pipes. If the risers aren't high enough, add the shortest possible temporary extension that will extend them above the ground level.

Open the valve to the full-open position, and let the water run for at least 5 minutes. The water should run down into the partially backfilled trenches; the water will cause the backfill dirt to settle around the pipes. Use a shovel to add dirt to the trenches as they start to settle. Poke the dirt gently to loosen any air bubbles. Once the entire valve circuit has been flushed, it is time to flush the individual risers. Cap all the risers except one. Slowly increase the water flow until a small geyser of 4 or 5 ft is coming from the riser. Continue flushing for at least 2 minutes. Repeat this process for each riser. When all the risers on the valve circuit have been flushed, repeat the entire process for the other valves, starting each time with a 5-minute flush of all the risers together.

The initial flush cycle with all the risers open serves two purposes: It provides water to settle the partial backfill, and it creates a high water demand (velocity) on the water supply, flushing out the sediment from the supply lines. Subsequent flushing of individual risers removes sand and debris from the irrigation system pipes and helps to identify any problems.

Installing Sprinklers

As soon as the pipes are thoroughly flushed, the sprinkler heads can be installed. To allow for edgers and other yard tools to operate without hitting the sprinklers, leave a border of 1 to 6 in between the perimeter heads and concrete edges. Try to get the riser worked into a position where the sprinkler stays where you want it without being supported, or at least with the least amount of support possible. If the sprinkler heads keep bending over at an angle or wobble a lot, try adding a 4-in length of pipe to the bottom inlet of the sprinkler with a TxT ell at the bottom of the pipe. Connect the riser to the ell. This essentially makes the sprinkler body longer and more stable.

Backfill around the sprinkler head, and compact the dirt around the body of the sprinkler. The soil level should be $^1/_4$ to $^1/_2$ in below the top of pop-up sprinklers. Shrub sprinklers on pipes above ground should be avoided where possible. Where they must be used, make them tall, and if possible, attach them to a large post, so that they are very visible and not a trip hazard. Never use shrub heads near sidewalks, driveways, or other pedestrian paths.

Automated Sprinkler Systems

Sprinklers are popular systems with homeowners and are prime candidates to control with an intelligent home. Programmable sprinklers have been around for years. These sprinkler systems are normally installed with an independent controller that the homeowner uses to preset the times during which one or more zones will be watered. One drawback of this type of system is that these controllers are limited by their own programming.

In situations such as parties, open houses, water-rationing days, or any event utilizing the yard for socializing, the owner may have to reprogram these independent controllers to prevent an unexpected drenching.

Some installations include an interface or controlled receptacle to the sprinkler system, providing remote control over the lawn sprinkler system. An interface adds event scheduling to the system's capability. Preprogrammed watering times can be changed or canceled by selecting an event such as "party" or "water ration" using a controller inside the home. These adjustments are generally short-lived (one or two days); the controller will automatically revert to the standard scheduling at the end of the event programming or when the homeowner places the system into the "normal" mode.

Just as homes can be divided into zones for HVAC, yards can be zoned for irrigation. A popular method for automating a sprinkler system is to use a relay interface device to control the 24-V valves. The interface provides a series of low-voltage dry contacts. This interface is usually placed into the circuit between the low-voltage power supply and the solenoid valves.

Connecting the relay to a whole-house automation controller allows interoperability with other automation devices. For example, on a daily basis home weather instruments measuring a wide range of weather conditions including various temperature zones, humidity, wind speed and direction, and precipitation can automatically set the watering schedule. Responding to the outside temperature for the current day, or a number of previous days, and factoring in the amount of rain within a given time, systems regulate the watering frequency from everyday to once every several days.

Many lawn sprinkler systems include their own moisture sensors, which shut off the watering if it rains or if adequate moisture is sensed. Some sensors are located aboveground, to detect rainwater only. Other sensors are buried in the ground to detect ground moisture. Both types typically function by opening or closing the ground circuit to the solenoid valves when moisture is detected.

Sprinkler systems often include their own dedicated controller. This device is commonly placed in a garage or other convenient location, and the homeowner can use it to set the watering times. In a home network, the use of a dedicated sprinkler controller offers advantages. The homeowner can use the dedicated controller to set watering times and can use other remote controls to disable it whenever the watering is not wanted. This method allows the lawn sprinkler system to remain active and operable if the automation controller fails or is out of service for changes or repair.

Probably the simplest method of accessing the sprinkler controller is by plugging the power supply for the solenoid valves into an electronic appliance switch or a controlled receptacle. Appliance switches offer the advantage of manually turning the system on or off from a wall switch conveniently located near the sprinkler controller. If that is not convenient, an interface can also be connected into the ground circuit with the power supply.

Watering Interior Plants

Living plants impact our lives in many ways. Research over the past 20 years has unlocked technical attributes of plants useful in helping us provide healthier home and work environments. We spend so much time indoors that the use of houseplants now takes on a new meaning. Filtering our air and brightening up the atmosphere are only two of the major contributions plants make to our homes. Automated plant care systems bring an advanced level of practicality to green use indoors.

Caring for interior plants is a labor-intensive task, particularly making sure each plant has the right amount of water. Plant care is a natural extension of home automation. The use of interior irrigation systems relieves the maintenance burden of plant care. While not meant to replace human plant care, these systems are capable of completely eliminating the watering tasks in the home. This is a convenience for homeowners and, when applied to the workplace, is an important cost-saving factor for office managers.

New technology has refined the exterior sprinkler and drip irrigation technologies that preceded it. The new technology is actually a precisely controlled flow system specially engineered for interior use. It permits complete building coverage with automated service, particularly for containerized plants in furnished, residential living areas and office workspace. It overcomes most of the drawbacks of self-watering containers. Techniques allow interior watering systems to be integrated into central systems of new construction or retrofitted into existing homes.

These systems are fully compatible with microprocessor-based energy management and other control systems. The interface is a simple one, requiring only a dedicated power receptacle, controlled by the master system. Solenoid valve controllers simply plug into the receptacle. Many configurations are available to accommodate a variety of needs. Some of these systems are available in two forms: low-pressure and high-pressure. The difference lies in the method of developing the dynamic forces that permit water to flow through the system. Low-pressure versions use a plastic reservoir that has a small pump integrally mounted. They are used when connections to a cold water pipe are not readily available. High-pressure systems are connected to the cold water plumbing and rely on the home's water pressure for flow.

Most systems share several common elements. Normally short pulses of water flow are used, enough to furnish the limited moisture requirements of indoor plants while providing the system safety and control. In average installations, small water applications are usually repeated twice daily, at regular intervals. The short cycle reduces tubing networks to minimum pressurization, providing an important safety feature.

Systems are designed to consistently maintain a user-selected, optimum level of moisture for each plant, providing a practical balance of moisture and oxygen around the roots. Frequent, gentle watering cycles are the preferred method of watering indoor plants since moisture has a chance to diffuse through the soil mass between irrigation cycles. Root rot and moisture stress are minimized. Less water is used as surface tensions don't have to be overcome in rewetting dried soil. Runoff and overflow problems are eliminated.

Controllers that power pumps or solenoid valves for very short periods are the hearts of the systems, determining when irrigation cycles occur and for how long. Generally, high-pressure systems use electronic controllers containing integral solenoid valves. A master control timer provides a power window at the dedicated power receptacle, just long enough for the controller to briefly activate the solenoid valve, causing flow to occur during that period. After the controller shuts down, the power window also shuts down. This arrangement provides an important safety feature. There is a double-tier control sequence, preventing runaway irrigation cycles and flooding. The first timed sequence is from the master controller; the second is from the solenoid valve controller's electronics.

Some systems are designed with remote control jacks that allow the homeowner to manually control flow with a thumb switch while adjusting emitters at planter locations. This is necessary during installation and for readjustments. Water pressure regulators reduce irrigation system pressure at the solenoid valve outlet, to 25 pounds per square inch (lb/in^2) or less, comprising another safety feature. They are the hydraulic equivalent of voltage regulators. Another common thread relates to the fact that these are central systems, with control centers mounted in convenient locations such as garages, laundry rooms, cabinets, or basements. The water distribution tubing network (seen in Fig. 10.5) is routed throughout the structure to the many existing planter locations (and potential planter locations). Tubing is installed to remote locations to service built-in planter pits, boxes, or freestanding container plants.

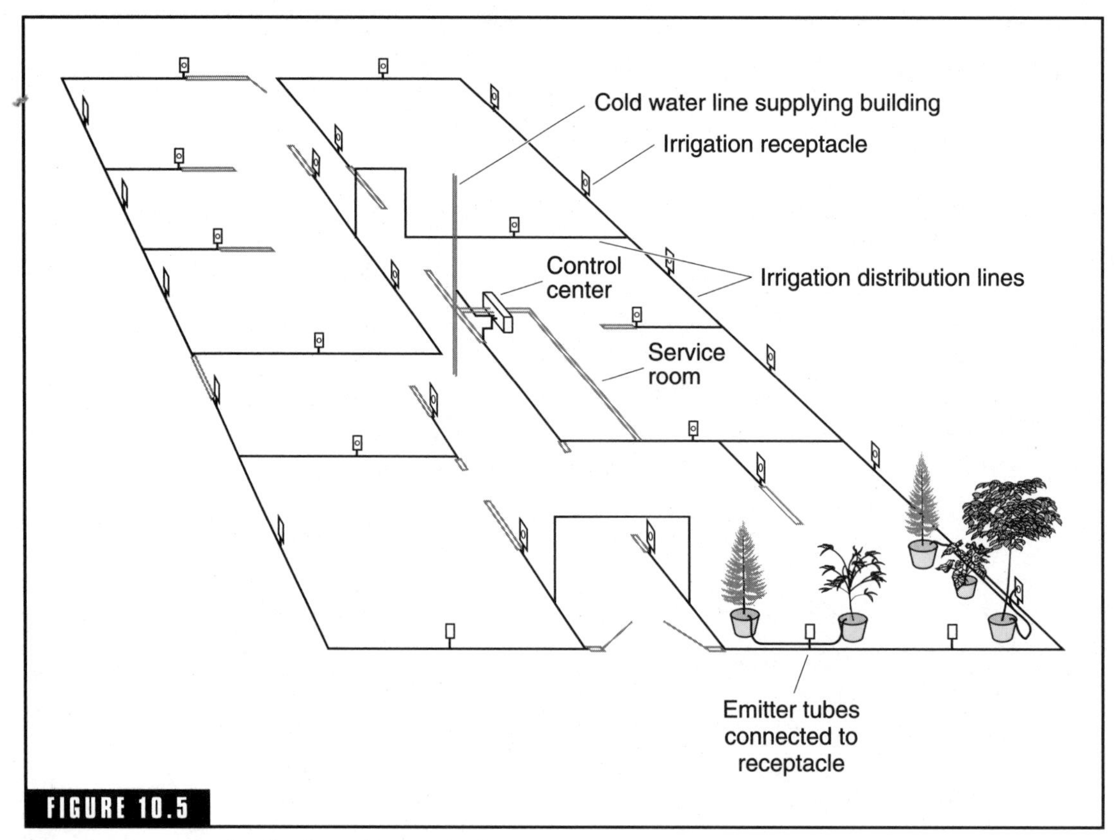

FIGURE 10.5

Example of a water distribution tubing network.

The routing is chosen for its aesthetic and technical practicality, hidden as much as possible, as are cable and wire runs. Tubing is generally flexible, small-diameter cross-linked polyethylene (PEX) or chlorinated polyvinyl chloride (CPVC), determined in large measure by local building codes. These are plumbing-grade materials. This type of tubing is easy to install, but as with everything, care must be taken. To eliminate errors, tubing systems are checked for leaks before partitions are sealed. Fittings are plastic, solvent-welded (CPVC) or brass slip-in fittings, secured with crimp rings (PEX), in much the same way as coaxial cable fittings are installed. Whenever possible, lines are routed through wall partitions, ceiling plenums, basements, attics, and crawl spaces. The best and most cost-effective installations are generally planned before construction, so that tubing runs can be designed into concrete floor slabs, wall partitions, and ceiling plenums. In this way, the irrigation system can be fully integrated into the building structure. Various flow control devices are installed in the tubing lines to control flow volume, pressure, and direction wherever necessary. Of these, check valves are the most useful. The tubing network must be filled with water at all times to eliminate the need for time-wasting air purges at the beginning of each irrigation cycle. Check valves provide that control.

A number of plants can be fed water from each emitter tube by branching them (with an adjustable emitter at each plant). Tubing branches of this type can traverse large rooms. Planter boxes, beds, pits, and shelves can be serviced with these systems as well. For the larger planter groupings, irrigation manifolds are used, terminating in emitter tubes for each plant, with the adjustable, minivalve emitter providing fine-tuned flow control. With proper planning, branches of the system can generally be extended to outdoor locations and to water-limited numbers of potted plants on patios and sun decks.

Pools and Spas

Some pools and spas include intelligent controls to automatically fill them and detect when a person has entered the water. Automatic filling can be accomplished with a solenoid valve and moisture sensor, which is connected to a plumbing source. These components are similar to the solenoid valves and moisture detectors used in sprinkler systems.

For homes with small children, sensors can be added to pools and spas to detect when someone enters the water. These sensors either are independent devices or more commonly are connected to the home's security system.

The most common intelligent controls used for pools and spas involve controlling the lights, pumps, and heaters. Pool lights can be controlled with electronic appliance switches similar to those used for indoor lighting. Heaters only turn on when the pump is on, so they are automatically controlled by the devices which control the pump. Pool and spa pumps are usually hardwired or plugged into a 220-V receptacle; electronic appliance switches or a controlled receptacle or module can be used to control these pumps. The switch offers the advantage of easy manual control near the pump location.

When switches or receptacles are installed, the homeowner can turn these devices on or off by remote control or by using an automation controller. For example, the pump can be set up to run for a predetermined time during the middle of the afternoon each day, to take maximum advantage of solar heating. The program can also create an entertainment scheme, automatically adjusting the pool and patio lights to preset levels. Another scheme may have the whole system shut down indefinitely when homeowners go on vacation. Spa controls are similar to pool controls, except that many spas include their own timers and control panels. An interface can be added to the circuits, parallel to the control panel wiring. This arrangement allows thehomeowner to turn the spa on or off by remote control or to have it automatically come on when an entertainment scene is activated.

Anti-Icing

Anti-icing equipment is often used for driveways, sidewalks, gutters, and roofs. There are two technologies which are commonly used with anti-icing systems: electrical and hydronic. Electrical deicers have an electric cable which is interlaced back and forth below the surface that is to be kept free of ice. Hydronic systems use a liquid antifreeze that is circulated inside tubing. Some hydronic systems use heat from the hot water tank, while others use independent boilers.

Both of these systems employ dedicated controllers to activate and deactivate the system. These controllers come with several types of sensors. Some have temperature sensors, which automatically turn the system on whenever the temperature falls below freezing. Others incorporate moisture sensors that activate the system only if both temperature and moisture conditions warrant it. Those which include the moisture sensor are generally recognized as more efficient.

Anti-icing controllers can be connected to home automation systems as lights are, through an electronic wall switch or controlled receptacle. Use of a wall switch provides both manual control and remote control. When an automation controller is used, it can be programmed to automatically function with the same flexibility as lights and pool equipment.

Convergence

Roughly defined, *convergence* refers to the merging of television and personal-computer technology to form a two-way experience instead of just a viewing experience. Also called *interactive TV* (*ITV*), convergence would change television programming and how homeowners watch it. ITV would allow homeowners to sit down in front of a television with a remote control, search for movies or an old television show, and watch it there and then. Viewers could order a CD while watching a concert on television—it might arrive in the mail or be downloaded immediately to the home's entertainment center.

Shopping in a virtual mall via the television would be done with the help of electronic catalogs; selected items would be delivered. Using computer technology, homeowners could see how new clothes or makeup might look on themselves by using a "virtual body." Interactive games can be played on televisions or downloaded into the games console of the entertainment center. News from local, national, or international sources can be called up whenever desired, free of traditional programming limitations.

Some of these features are available today, and others plus much more are currently being tested by a group of media, cable television, telephone, and computer companies. The base necessary to implement this technology already exists at a sufficient level to allow limited trials to be run. However, very significant developments are required in many areas before it will be commercially viable to extend the full array of features to the public. To provide any of the services mentioned, the following components are required:

Content: Content means any form of source material—movies, games, news, images, sounds, etc.—which will appear on the user's television or PC screen.

Compression capabilities: Most of the services can only be achieved effectively by using digital technology; systems are required to convert analog signals to digital signals and to store them in a highly compressed format.

Storage hierarchy and control system: Even compressed videos require enormous amounts of storage space; the control system must be able to service all the requests coming in.

Transmission system: High-speed links are required to deliver the vast amounts of information used in the system in a timely manner.

Return path: In a fully interactive system there needs to be a signal going from the user to the control system carrying the user's requests.

Set-top box: An addressable communications box is needed to decode the signals as they arrive at the television. Depending on the system used, it may also need to perform functions such as decompression of the digital signal.

Remote control and navigation system: Users need a friendly interface to find their way through all the services offered and communicate their requirements to the central control system.

Subscriber management: Sophisticated systems for administration, billing, and encryption will be required to ensure that the users pay for the services they use and that copyrights are preserved. Forms of convergence technology available today include set-top boxes that are connected to televisions. These boxes, when connected to a phone line, provide access to the Internet over the television. For example, WebTV has just recently introduced a second generation of its television-based Internet access service that allows users to view television programming and the Web at the same time. WebTV is now offering a system dubbed WebTV Plus that integrates a cable-ready TV tuner and a 1.1-gigabyte (1.1-Gbyte) hard

> drive into its set-top box. The WebTV Plus network service features TV crossover links, which complement and enhance TV programs by providing quick and easy access to integrated Web sites that are directly related to popular TV programs. The WebTV Plus receiver detects Web links and embedded data in video broadcasts. The system notifies viewers of these additional data via a small icon that appears on the television screens.
>
> TV crossover links give audiences a variety of entertaining possibilities to choose from. They may find breaking news or gossip about their favorite stars or a synopsis of past episodes. They might engage in on-line chatting with fans or stars, or offer their feedback through instantaneous polls and votes. By giving viewers more ways to interact with their television programming, TV crossover links also provide broadcasters and advertisers with a powerful new medium to connect more closely and efficiently with their audiences. Leading content providers such as Discovery, E! Online, PBS, Warner Bros. Online, and MSNBC will be taking advantage of Web-based content to enhance their television programming.

Set-top boxes are easily installed (usually by the homeowner), but having the home prewired with coaxial cable and telephone wire makes the job much easier. Main cable leads are connected to the set-top box with a separate coaxial cable line connecting the box to the television. A telephone line is also connected to the set-top box to provide modem access.

Applications

Maybe NFL football (an NBC interactive offering) with its statistics-hungry fans will be a driving force pushing ITV forward. On the other hand, maybe it will be a more traditional PC feature, like e-mail on the TV set. E-mail would take on new meaning; instead of old-fashioned keyboard e-mail, ITV could accommodate streaming video. A camcorder is connected to the system, and live video of the family can be flashed to friends and relatives. The prospect of computer and television merging into one entity has caught the attention of many large businesses. General Electric, Viacom, Time Warner, Philips, Sony, Intel, Microsoft, and a host of other corporations are scrambling to secure their piece of this new market.

In theory, programmers will be able to publish Web sites designed to augment their television shows. In addition, companies such as Microsoft could download video programming for homeowners to view later. Some of the technology behind convergence, such as the software that allows video to run on a computer screen, has been around for years. The biggest barrier to interactive television has been, and continues to be, the lack of a conduit into the home with enough bandwidth to carry vast streams of digital data back and forth. The two media are converging now partly due to the following factors:

- The increased power of microprocessors
- The cost of memory chips dropping significantly
- The number of home-based personal computers
- The popularity of the Internet
- The amount of digital information now available

Intel Corporation's universal serial bus facilitates small digital devices such as modems, digital cameras, and computer peripherals being plugged directly into the computer's serial or parallel port. Once plugged in, these devices are automatically configured by Microsoft Windows 98. The next generation of convergence products (available in a year or so) will handle bigger toys. Digital TV sets, digital versatile disc players, and digital CD changers will all be convergence items.

Transmission Types

Video transmissions not only must be at a sufficiently high rate but also must be delivered isochronously. There are a number of different methods of transmission: twisted-pair wire, coaxial cable, fiber-optic cable, satellite, and microwave. Different systems have different limitations.

Satellite

Satellite is really a wireless broadcast medium. Although it will handle many interesting new applications in the future, there are some interesting challenges for fully interactive systems in which every user requires a different signal to be sent to her or his television. Another key problem is the latency (delay) inherent in a satellite-based system.

Extensive work is being done in this area with new methods of packaging the data to reduce the delays during an interactive conversation. Satellite transmission is also a one-way system providing no possibility of a return path. However, telephones provide a very simple return path: dialing up the control center and using touch-tone codes to communicate with the control system.

Twisted Pair

This most common wired system is present in millions of telephone lines going to houses. It is also the most limited in its bandwidth. However, the upgrading of the backbone networks has meant that the limitations are often in only the "last mile" as the wire enters the house, and improvements in technology enable the necessary speeds to be transmitted over that last mile. ADSL-1 (Asymmetric Digital Subscriber Loop) allows 1.5 Mbits/s (that is, MPEG-1) to be transmitted into the home and provides a low-speed (64 Kbits/s) return path suitable for most interactive applications. ADSL-2 should allow MPEG-2 transmissions at up to 6 Mbits/s. Twisted-pair wire is not fast enough to allow the analog transmissions which could take advantage of cheap set-top boxes; but since the future is undoubtedly digital, this is only a short-term limitation. A more likely limiting factor on the development is the much higher speed available on coaxial and fiber-optic cables—there may be insufficient interest in twisted-pair wire on a worldwide basis to ensure its success.

Coaxial and Fiber-Optic Cable

Coaxial cable can provide 100 channels, each of which is effectively a 36 Mbits/s pipe. These pipes can be further broken down into twelve 3 Mbits/s MPEG-2 digital television channels, thus giving a total of 1200 channels (plus spare capacity for control and management) as opposed to one on a twisted pair. (There are many variations on this calculation, but all indicate an enormous number of channels.) Likewise, a fiber-optic cable can provide up to 150,000 times the capacity of a twisted pair. Since current cable systems are used mainly for broadcasting signals now, to become fully interactive requires the introduction of switching systems to enable true one-to-one transmissions to take place. The telephone systems do have the advantage here of being designed with such switching systems already.

Another difference between twisted pair and coaxial cable is the status of the existing networks. In the United Kingdom, a single telephone company services virtually every home while less than 10 percent of homes are wired for cable television. In the United States, more than 90 percent of homes are within easy reach of a cable television system (more than 60 percent are already connected), but the telephone network is owned by a number of competing companies.

Each country has its own profile of existing services, complexity of local regulations, and government interest in the various industries which are involved in competing for the lion's share of these future marketplaces. This adds further social, political, regulatory, and economic complexity to the technological challenges. All these transmission systems are being used in trials today; whether any will become standard is hard to say. In fact, the end result is likely to be a combination of many technologies, deriving benefits by using the most appropriate technology in the most appropriate place.

Return Path

Most systems are already capable of providing a return path, and the telephone system is always available as an alternative. The return path for most interactive television applications does not need to be very fast (64 Kbits/s is quite adequate). The return path is used to transmit short bursts of control information, as opposed to the supply side sending large segments of media to the home along the main path. Although sufficient for movies, shopping, and standard entertainment fare, this return path is not sufficient for two-way video applications such as video conferencing or video stream e-mail.

Remote Control and Navigation Systems

The navigation system, which allows a user to select a service and specify the particular aspect of that service required, is a critical part of a successful system. For example, the user must be able, with a simple remote control, to select a film, then search by different criteria until locating the particular film desired.

To surf through 500 channels would take about 43 minutes, by which time new programs would be starting. Studies have shown that of 50 channels, homeowners generally use only about 7. An intelligent set-top box can help eliminate time-consuming channel surfing. These boxes can be programmed with the personal preferences of the family, allowing a list of matching shows to be presented on screen.

New Products and Developments

Some home automation installations may require simply wiring the home for current and future needs. Other installations may require actually installing equipment such as controllers, sensors, thermostats, and entertainment equipment. It is therefore helpful to be at least somewhat familiar with some of the latest technology being marketed. Remember, the most important aspect of the job is to understand the needs and desires of the homeowners; part of that understanding comes from having a working knowledge of the equipment.

VocalNet

Internet phone products have proliferated over the past few years. However, new products promise to take voice communication over the Internet to the next level. One of these products is VocalNet by Inter-Tel. This product consists of a dedicated server that connects between a telephone system and an Internet connection, converting voice information to Internet protocol data packets and transmitting them in real time over the Internet. This dramatically reduces the bandwidth necessary to transmit voice data. The other end will need to have VocalNet as well.

C-Phone Home

C-Phone Home from C-Phone Corporation is a video telephone that uses standard telephone lines for making videophone calls. The unit is a completely self-contained TV set-top box that has a wireless user interface. The product includes a camera, a built-in 33.6 Kbits/s modem, and a wireless remote control. A standard television is used to present the audio and video of the person being called as well as all text menus used to operate the system. A built-in microphone, in conjunction with the TV's speaker, forms a full-duplex speakerphone. The speakerphone can be used for standard calls as well as videophone calls.

QuickSilver Hydra

TeleSite USA, Inc., has released QuickSilver Hydra, a phone line video transmission system that allows the operator to view up to four separate remote sites simultaneously on one computer screen. The system supports an unlimited number of sites, with each site having up to 48 cameras and up to 12 cameras with pan, tilt, and zoom capabilities. If it is used with a speakerphone at the remote site, audio and video can be transmitted together on a single telephone line. QuickSilver transmitters can store up to 6000 frames at the remote site without the use of a PC, allowing images captured prior to transmission to be replayed.

HomeVision-PC

HomeVision-PC is a low-cost version of the Custom Solutions, Inc., flagship home automation controller. The new controller provides two-way X-10 (all 256 addresses), two-way learning infrared (255 signals), 24 I/O ports, and video output screens displayed on a TV. It also includes system scheduling and "if-then" programming, graphical user interfaces, voice recognition and response, and telephone control. The HomeVision-PC does require an operating PC. Packaged in a metal enclosure, HomeVision-PC connects to the computer's serial port and comes complete, including Windows-based software.

Home Director

IBM has released two new lines of Aptiva computers featuring its new Voice-Enabled Home Director home automation controller. Two models, the Aptiva L5H and S6H, will feature the Voice Enabled Home Director as a standard feature, while the S62, S6S, L31, and L61 will offer it as an option. This new version of Home Director will use IBM's new VoiceType Simply Speaking software to allow voice control of the Home Director software. The new Aptivas also feature IBM's new Scrollpoint mouse. This mouse features a *pointer* control that allows 360° scrolling without the need to move to a scroll bar.

Leviton's MOS

Leviton Telcom has expanded its line of *multimedia outlet system* (*MOS*) inserts to include S-video and RCA modules, the two most widely used high-quality, high-definition audio/video connections.

The module features a "no soldering required" design using true direct-termination connections. This feature delivers unobstructed signal reception and higher performance. Installation is also less time-consuming and less labor-intensive because raw S-video and RCA cabling can be terminated by using insulation displacement connectors (S-video) or screw terminations (RCA) to the printed-circuit boards built right into the modules.

Denon DVD-3000

With the proliferation of digital disc media such as DVD, audio CDs, and video CDs, equipment cabinets can become crowded. Denon's DVD-3000 player can read all the above formats. It is also DTS digital surround ready, and it features Dolby Digital (AC-3) decoding onboard. Six-channel analog outputs and video outputs including two for composite video, one for S-video, and one set for component video outputs (yellow, blue-yellow, red-yellow) are all included.

NAD Distribution Preamp

NAD has introduced the model 911 distribution preamplifier, a preamplifier capable of routing to as many as six outputs from six different sources, each with independent level controls. The insert line out on model 911 allows daisy chaining of the units, making the number of channels supported virtually unlimited. A microphone input gives the preamplifier the capabilities of a paging system, and a microphone or line-level signal can be inserted into any of the channels. Each channel has its own on/off switch, allowing full control of how messages are heard.

Speakercraft "WavePlane Technology"

Speakercraft has introduced the 6.1 DT speaker, which is capable of reproducing stereo sound from one speaker, with new WavePlane Technology for improved sound dispersion. In whole-house audio installations such as laundry rooms and dressing areas, audio either is a monoaural feed from one side of the source or requires installation of two speakers in a tight space. The 6.1 DT uses a pair of polyamide dome tweeters with a 95 percent solid surface and a 6-in woofer, combined with specially designed louvers that channel the sound waves. Installation is simple—just plug both the left and right speaker cables into the push terminals on the back of the speaker, and turn the four locking screws to hold the speaker securely in place.

Addressing the needs of the home theater user, Speakercraft has also introduced two new cabinet speakers, designed for use as center, front left, or front right speakers in a home theater audio system. The 6.5LCR and 6.1LCR both feature a 1-in silk dome tweeter that can be pivoted to any spot in the listening area. Also part of the speakers are two 6.5-in video-shielded woofers, for clean extended midrange and bass. The woofers in the 6.5LCR are mineral-filled with butyl rubber surrounds. The 6.5LCR also has front adjustable bass and treble switches.

Z-Man Audio Signal Enhancer

Music from digital sources is often cold and overly bright. The Z-Man Audio Signal Enhancer is designed to bring back the natural warmth of musical instruments and voices, restoring the full sound originally recorded. The unit installs easily between the source, such as a CD player, and the amplifier or preamplifier. The manufacturer, Z-Man Corporation, claims the device can enhance the signal of virtually any source.

Solar Light at Night

Alpan's Pagoda Garden Light is different from standard walkway lights, as it uses Alpan's Monterey Series power supply and timer. This system absorbs solar energy and stores it for use at a later time. In addition, the Pagoda uses 9-W compact fluorescent bulbs that put out as much light as a standard 35-W incandescent light, while saving energy. The lights have built-in photosensors and timers for automatic on/off and dusk/dawn switching.

Summary

The technologies and devices discussed in this chapter and throughout this book are just the beginning of home automation. As homeowners become more familiar and comfortable with home automation, every aspect of life and work around the house will eventually be touched by some form of automation.

Most homeowners will find that there are certain areas where great control flexibility is desired whereas most others need minimal capability. You should key on the homeowner's needs and help differentiate needs from desires. If budget constraints keep homeowners from having all their home automation needs met up front, the full-house prewiring will accommodate all foreseeable subsystems and controls. Individual subsystems may be upgraded and/or added as the budget permits.

The degree of automation and complexity of the individual subsystems will determine the level of expertise required in both the systems integrator as well as the general contractor and subcontractors. The role of the automation designer-installer is to facilitate communication between the general contractor and the homeowner as well as between the general contractor and the affected subcontractors. In extensive automation systems, it is ideal for the general contractor to have knowledge of and experience with automation systems.

Notes

Notes

Notes

APPENDIX

A

Data Communications and Cable Manufacturers

AMP:

Addr.: AMP
Harrisburg, PA 17105-3608
Tel.: (800) 722-1111
(800) 245-4356 (Fax-back service)
(905) 470-4425 Canada
(617) 270-3774 (Fax-back service, Canada)

Anixter:

(An international cable products distributor; see *Anixter Cabling Systems Catalog*)
Addr.: Anixter, Inc.
4711 Golf Road
Skokie, IL 60076
Tel.: (708) 677-2600
(800) 323-8167
(800) 361-0250 Canada
32-3-457-3570 Europe
44-81-561-8118 United Kingdom
65-756-7011 Singapore

ANSI:

Addr.: American National Standards Institute
11 W. 42d St., 13th floor
New York, NY 10036
Tel.: (212) 642-4900

AT&T Canada:

Addr.: Network Cables Division
1255 Route Transcanadienne
Dorval, QC H3P 2V4
Tel.: (514) 421-8213
Fax: (514) 421-8224

AT&T documents:

Addr.: AT&T Customer Information Center
Order Entry
2855 N. Franklin Road
Indianapolis, IN 46219

Tel.: (800) 432-6600
(800) 255-1242 Canada
(317) 352-8557 International
Fax: (317) 352-8484

Belden Wire & Cable:

Addr.: Belden Wire & Cable
P.O. Box 1980
Richmond, IN 47375
Tel.: (317) 983-5200

Bell Canada:

Addr.: Bell Canada
Building Network Design
Floor 2, 2 Fieldway Road
Etobicoke, Ontario
Canada M8Z 3L2
Tel.: (416) 234-4223
Fax: (416) 236-3033

Bell Communications Research (Bellcore):

Addr.: Bellcore Customer Service
60 New England Ave.
Piscataway, NJ 08854
Tel.: (800) 521-2673
Fax: (908) 336-2559

Berk-Tek: (Copper and fiber-optic cable)

Addr.: Berk-Tek
312 White Oak Rd.
New Holland, PA 17557
Tel.: (717) 354-6200, (800) BERK-TEK
Fax: (717) 354-7944

BICSI:

[A telecommunications cabling professional association. Offers education and administers the RCDD (Registered Communications Distribution Designer) certification]

Addr.: Building Industries Consulting Service International
10500 University Center Drive, Ste. 100
Tampa, FL 33612-6415
Tel.: (813) 979-1991, (800) BICSI-05
Fax: (813) 971-4311

Black Box

Black Box Catalog: The Source for Connectivity

Addr.: Black Box Corporation
1000 Park Drive
Lawrence, PA 15055-1018
Tel.: (724) 746-5500
Fax: (724) 746-0746
Inet: info@blackbox.com

CABA:

Addr.: Canadian Automated Buildings Association
M-20, 1200 Montreal Rd.
Ottawa, ON K1A 0R6
Tel.: (613) 990-7407
Fax: (613) 954-5984

CableTalk:

(Racks and physical cable management)

Addr.: CableTalk
18 Chelsea Lane
Brampton, ON L6T 3Y4
Tel.: (800) 267-7282, (905) 791-9123
Fax: (905) 791-9126

Cabling Business:

Addr.: Cabling Business Magazine
12035 Shiloh Road, Ste. 350
Dallas, TX 75228
Tel.: (214) 328-1717
Fax: (214) 319-6077

Cabling Installation & Maintenance Magazine:

Addr.: Cabling Installation & Maintenance Editorial Offices
One Technology Park Dr.
P.O. Box 992
Westford, MA 01886
Tel.: (508) 692-0700
Subscriptions:
Tel.: (918) 832-9349
Fax: (918) 832-9295

CCITT: See ITU.

Comm/Scope, Inc.

Addr.: P.O. Box 1729
Hickory, NC 28603
Tel.: (800) 982-1708, (704) 324-2200
Fax: (704) 328-3400

Corning/Siecor:

Addr.: Siecor
489 Siecor Park
P.O. Box 489
Hickory, NC 28603-0489
Corning Optical Fiber Information Center
(800) 743-2675
Guidelines—publication/newsletter on fiber technology
Fax: (828) 327-5891
Inet: http://www.siecor.com

CSA:

Addr.: Canadian Standards Association
178 Rexdale Blvd.
Rexdale, Ont.
Canada M9W 1R3
Tel.: (416) 747-4000
Document orders: (416) 747-4044
Fax: (416) 747-2475

EIA:

Addr.: EIA Standards Sales Office
2001 Pennsylvania Ave., N.W.
Washington, DC 20006
Tel.: (202) 457-4966

GED:

Addr.: Global Engineering Documents
1990 M Street W, Ste. 400
Washington, DC 20036
Tel.: (800) 854-7179, United States and Canada
(202) 429-2860 International
(714) 261-1455 International
Fax: (317) 352-8484

Global Engineering Documents (West Coast)
Addr.: 2805 McGaw Ave.
Irvine, CA 92714
Tel.: (800) 854-7179

Graybar:

(An international cable products distributor)
1-800-825-5517
Tel.: (519) 576-4050 in Ontario
Fax: (519) 576-2402

Hubbell:

Addr.: Hubbell Premise Wiring Inc.
14 Lords Hill Rd.
Stonington, CT 06378
Tel.: (203) 535-8326
Fax: (203) 535-8328

IEC:

Addr.: International Electrotechnical Commission
Rue de Varembre, Case Postale 131,3
CH-1211
Geneva 20, Switzerland

Addr.: U.S. National Committee of the IEC
ANSI
11 West 42nd Street, 13th Floor
New York, NY 10036
Tel: (212) 642-4900
Inet: http://www.iec.ch
http://www.ansi.org

ISO:

Addr.: International Organization for Standardization
1, Rue de Varembre, Case Postale 56
CH-1211
Geneva 20, Switzerland
Tel.: 41-22-34-12-40

Mod-Tap:

(Cable and equipment supplier)
Addr.: Mod-Tap
285 Ayer Rd., P.O. Box 706
Harvard, MA 01451
Tel.: (508) 772-5630
Fax: (508) 772-2011

NFPA:

[U.S. National Electrical Code (NEC) and other documents]
Addr.: National Fire Protection Association
One Battery March Park, P.O. Box 9146
Quincy, MA 02269-9959
Tel.: (800) 344-3555
Fax: (617) 984-7057

NIST:

Addr.: U.S. Department of Commerce
National Institute of Standards and Technology
100 Bureau Drive
Gaithersburg, MD 20899-0001
Tel: (301) 975-NIST

NIUF:

Addr.: North American ISDN Users Forum
NIUF Secretariat
National Institute of Standards and Technology
Bldg. 223, Room B364
Gaithersburg, MD 20899
Tel.: (301) 975-2937
Fax: (301) 926-9675
Inet: sara@isdn.ncsl.nist.gov

Northern Telecom:

(Cable and physical network products)
Addr.: Northern Telecom
Business Networks Division
105 Boulevard Laurentien
St. Laurent, QC H4N 2M3
Tel.: (514) 744-8693, (800) 262-9334
Fax: (514) 744-8644

NRC of Canada:

Addr.: Client Services
Institute for Research in Construction
National Research Council of Canada
Ottawa, ON K1A 0R6
Tel.: (613) 993-2463
Fax: (613) 952-7673

NTIS:

Addr.: U.S. Department of Commerce
National Technical Information Service
5285 Port Royal Rd.
Springfield, VA 22161

Tel.: (703) 487-4650
(800) 336-4700, rush orders
Fax: (703) 321-8547

Ortronics:

Addr.: Ortronics
595 Greenhaven Rd.
Pawcatuck, CT 06379
Tel.: (203) 599-1760
Fax: (203) 599-1774

RCDD: See BICSI.

Saunders Telecom:

(Racks, trays, and accessories)
Addr.: Saunders Telecom
8520 Wellsford Place
Santa Fe Springs, CA
Tel.: (800) 927-3595
Fax: (310) 698-6510

SCC:

Addr.: Standards Council of Canada
1200-45 O'Connor St.
Ottawa, Ontario Canada K1P 6N7
Tel.: (613) 238-3222
Fax: (613) 995-4564

Siecor:

Addr.: Siecor
489 Siecor Park, P.O. Box 489
Hickory, NC 28603-0489
Tel.: (704) 327-5000
Fax: (704) 327-5973

Siemon:

(Cabling system supplier)
Addr.: The Siemon Co.
76 Westbury Park Rd.
Watertown, CT 06795
Tel.: (203) 274-2523
Fax: (203) 945-4225

TIA:

Addr.: Telecommunications Industries Association
2500 Wilson Boulevard, Ste. 300
Arlington, VA 22201
Tel.: (703) 907-7700
Fax: (703) 907-7727

UL:

(Underwriters Laboratories documents)
Addr.: Underwriters Laboratories Inc.
333 Pfingsten Road
Northbrook, IL 60062-2096
Tel.: (800) 676-9473 (Canada, east coast United States)
(800) 786-9473 (Canada, west coast United States)
(708) 272-8800 International
Fax: (708) 272-8129
Inet: 0002543343@mcimail.com

APPENDIX

B

Manufacturers of Security Components and Systems

Manufacturer	Products
ABM	Central station automation software
ADEMCO	Magnetic contacts, detectors, control panels
Advanced Technology Video	Video multiplexers, quads
Aiphone	Intercom systems, video entry systems, and nurse call systems
AlarmSaf	Power supplies and relays
Aleph	Magnetic contacts, photoelectric beams, motion detectors
Alpha-Comm	Intercom, nurse call, and apartment-entry systems
Altronix	Relays, timers, and power supplies

Manufacturer	Products
Ansul	Fire detection and suppression systems
ATC Frost	Transformers
C&K	Glass break detectors, motion sensors, control panels, magnetic contacts, and audible devices
Caddx	Control panels, glass break detectors
Cansec	Access control systems
Casi-Rusco	Access control systems
Channel Plus	Video modulators and distribution systems
Checkpoint Systems	Electronic article surveillance
Corby	Access control systems
Dedicated Micros	Video multiplexers, transmission and storage
Deltavision	Power supplies, CCTV cameras and peripherals, water detectors
Detection Systems	Smoke detectors, motion detectors
Detex	Access control locks
Door King	Access control systems
DSC	Digital Security Controls makes detectors, controls, smoke alarms, and accessories
Edwards	Fire controls and peripherals
Elk Products	Sirens, installation peripherals, telephone paging amplifiers
Europlex	Control panels
FBII	Control panels, relabeled peripherals from ADEMCO
Fiber Options	Fiber-optic transmission systems
Fire Lite	Fire controls and accessories
FM Systems	Signal-handling equipment
GRI	Magnetic contacts and accessories

Manufacturer	Products
Guardall	Access control
Gyyr	Time-lapse video recorders and video handling equipment
HAI	Building automation controllers
HID	Reader technology for access control
Hirsch Electronics	High-security access controllers
Hitachi	Cameras, video recorders, and monitors
Identicator	Electronic fingerprinting systems
ICI	Access controls
IEI	Electronic keypads and stand-alone access management
ITI	Wireless security controls and detectors
ITT Nightvision	Nightvision products
Kantech	Access control panels, readers, badging systems, and peripherals
Keri Systems	Stand-alone proximity access control systems
LRC Electronics	CCTV training products
Labor $aving Devices	Installation tools and accessories
Litton PolyScientific	Nightvision and perimeter intrusion sensors
Master Video	CCTV and CATV modulators, character generators
Menvier	Control panels and voice dialers
Micro Key Software	Central station automation software
Mircom	Apartment intercom, intercom, nurse call, and fire products
Napco	Motion detectors, control panels, access control
Northern Computer	Access management and badging systems

Manufacturer	Products
Optex	PIR detectors, photobeams, control panels
Opticom	Covert and miniature cameras and accessories
Osborne-Hoffman	Central station receivers
Pelco	Camera domes, housings, pan/tilts, and control systems
Philips	CCTV cameras, enclosures, switchers, quad displays, and recorders
Potter	Sprinkler flow switches
Pulnix	Photobeam and motion detectors
Radionics	Fire controls, security controls, and access controls
Rainbow	Closed-circuit television lenses
Rokonet	PIR detectors, glass break detectors, and add-on wireless products
Rutherford Controls	Stand-alone access control systems, door strikes, magnetic locks, access peripherals
Sanyo	Cameras, recorders, and monitors
SecuraKey	Access control systems
Securitron	Magnetic locks, egress devices, and access control peripherals
Sensormatic	Quad displays, multiplexers, video transmission, multiplex control equipment
Sentex	Access control and telephone entry systems
Sentrol	Control panels, smoke detectors, magnetic contacts, motion sensors, glass break detectors, plastic molding
Silent Knight	Control panels, fire controls
Silent Witness	Weatherproof, compact cameras
SIMS	Central station automation software

Manufacturer	Products
SIS	Central station automation software
SONY	Closed-circuit cameras, VHS time-lapse recorders, monitors, video printers
STI	Tamper-proof pull-station covers and accessories
Sur-Gard	Central station receivers, video transmission
System Sensor	Smoke detectors, signaling appliances, sprinkler devices
Tamron	CCTV lenses
Telexis	High-end video transmission applications
Trango Systems	Wireless FM video systems
Ultrak	Cameras, CCTV accessories
VCR Inc.	Video capture accessories
Videolarm	Camera housings and enclosures
Visions Televideo Technologies	Video transmission
Visonic	Motion detectors, glass detectors, wireless peripherals, voice activation
Visual Methods Inc.	Video access control
Von Duprin	Door-strike and exit devices
Watec	Closed-circuit cameras and remote positioning accessories
Wheelock	Fire-signaling appliances, paging amplifiers, and speakers
Winland	Temperature and water sensors
Winsted	Consoles, racking, and security furniture
X-10 Pro	Power line carrier automation devices
Yuasa/Exide	Backup batteries

APPENDIX

C

Conversion Tables

Metric to U.S. System Conversions, Calculations, Equations, and Formulas

Millimeters (mm) × 0.03937 = inches (in)

Centimeters (cm) × 0.3937 = inches (in)

Meters (m) × 39.37 = inches (in)

Meters (m) × 3.281 = feet (ft)

Meters (m) × 1.094 = yards (yds)

Kilometers (km) × 0.62137 = miles (mi)

Kilometers (km) × 3280.87 = feet (ft)

Liters (L) × 0.2642 = gallons (U.S.) (gal)

Liters (L) × 0.0353 = cubic feet (ft^3)

Kilograms per cubic centimeter (kg/cm^3) × 14.223 = pounds per square inch (lb/in^2, or psi)

U.S. System to Metric Conversions, Formulas, Calculations, and Equations

Inches (in) × 25.4 = millimeters (mm)

Inches (in) × 2.54 = centimeters (cm)

Inches (in) × 0.0254 = meters (m)

Feet (ft) × 0.3048 = meters (m)

Yards (yd) × 0.9144 = meters (m)

Miles (mi) × 1.6093 = kilometers (km)

Feet (ft) × 0.0003048 = kilometers (km)

Gallons (gal) × 3.78 = liters (L)

Cubic feet (ft^3) × 28.316 = liters (L)

Pounds per square inch (lb/in^2, or psi) × 0.0703 = kilograms per cubic centimeter (kg/cm^3)

Area and Distance Conversions, Equations, Calculations, and Formulas

Acres × 43,560 = square feet (ft^2)

Inches (in) × 0.0833 = feet (ft)

Feet (ft) × 12 = inches (in)

Square miles (mi^2) ("sections") × 640 = acres

Miles (mi) × 5280 = feet (ft)

Circumference of circle × 0.3183 = diameter of circle

Diameter of circle × 3.14 = circumference of circle

Diameter squared × 0.7854 = area of circle

Radius squared × 3.14 = area of circle

Water Pressure Equations, Conversions, Formulas, and Calculations

Feet-head (ft hd) × 0.433 = pounds per square inch (lb/in^2, or psi)

Pounds per square inch (lb/in^2) × 2.31 = feet head (ft hd)

Meters head (m hd) × 3.28 = feet head (ft hd)

Feet head (ft hd) × 0.3049 = meters head (m hd)

Flow and Water Volume Formulas, Conversions, Calculations, and Equations

U.S. gallons per minute (gal/min, or gpm) × 0.1337 = cubic feet per minute (ft^3/min)

Cubic feet per minute (ft^3/min) × 7.48 = U.S. gallons per minute (gal/min)

Cubic feet per second (ft^3/s) × 448.8 = U.S. gallons per minute (gal/min)

U.S. gallons per minute (gal/min) × 0.00223 = cubic feet per second (ft^3/s)

Acre-inches per hour (acre · in/h) × 453 = U.S. gallons per minute (gal/min)

British Imperial gallons × 1.201 = U.S. gallons (gal)

U.S. gallons (gal) × 0.833 = British Imperial gallons

Acre-feet (acre · ft) × 325,850 = U.S. gallons (gal)

Acre-inches (acre · in) × 27,154 = U.S. gallons (gal)

Velocity in feet per second = 144 × (gpm/inside diameter of pipe, squared)

Q = AV (quantity = area × velocity), the basic equation of water flow; example: quantity in feet per second = square feet of area × feet per second velocity

1 inch (in) of water depth = 0.62 gallon per square foot (gal/ft^2) of area

Pump Calculations, Conversions, Equations, and Formulas

The following formulas assume 55 percent pump efficiency (the standard assumption).

Horsepower is brake horsepower for an electric motor. Do not use for fuel-powered pump engines.

gpm = (horsepower × 2178)/feet-head

Feet-head = (2178 × horsepower)/gpm

Efficiency of pump = (gpm × feet-head)/(horsepower × 3960)

Miscellaneous Irrigation Formulas, Conversions, Equations, and Calculations

Note: Head spacing is width between sprinklers times length between sprinklers.

Precipitation rate for square sprinkler spacing:

$$\frac{\text{gpm of full-circle sprinkler} \times 96.3}{\text{head spacing squared}}$$

Precipitation rate for triangular sprinkler spacing:

$$\frac{\text{gpm of full-circle sprinkler} \times 96.3}{\text{head spacing squared} \times 0.866)}$$

GLOSSARY

absorptance (formerly absorption factor) Ratio of the absorbed radiant or luminous flux to the incident flux.

absorption Conversion of radiant energy to a different form by interaction with matter.

accommodation Focal adjustment of the eye, generally spontaneous, made for the purpose of obtaining maximum visual acuity at various distances.

acrylic fiber Fiber consisting of an inner acrylic plastic core coated with a thin cladding of a fluorinated resin. This material is more durable and more lightweight than glass fiber, can be bent to a tighter radius than can glass fiber, and can be easily field-terminated.

adaptation The process by which the state of the visual system is modified according to the luminances or the color stimuli presented to it.

air-handling unit A type of heating and/or cooling distribution equipment that channels warm or cool air to different parts of a building. This process of channeling the conditioned air often involves drawing air over heating or cooling coils and forcing the air from a central location through ducts or air-handling units. Air-handling units are hidden in the walls or ceilings, where they use steam or hot water to heat or chill water to cool the air inside the ductwork. (See **duct**.)

AC-3 An eight-channel sound format used in commercial movie theaters. Only six channels are used, and the sound is run off CDs. The supposed followup for home theater is DTS Coherent Acoustics.

anamorphically squeezed This process, which is used on few laserdiscs, a few DVDs, and even fewer TV broadcasts, is employed to achieve a wide-screen image, where the image is considerably wider than standard NTSC fare, once it is "unsqueezed." The wider image is squeezed into the skinnier aspect ratio, which is usually the NTSC standard of 4:3/1.33:1. Unsqueezing can be done with a "stretching circuit" in the TV. The end result (if left unsqueezed) is a picture with really skinny objects. Another option which has less detail but is more widely used is letterboxing the picture.

application sharing A feature of many document conferencing packages that lets a pair of users on different systems simultaneously use an application that resides on only one of the machines.

aspect ratio The ratio of the width of a picture to the height. The NTSC standard is 4:3. The current HDTV standard is 16:9, or 1.78:1. Modern movies range from 1.66:1 to 2.4:1. By far the most common are 1.85:1 and 2.35:1.

axial mode-allows The highest light output for fiber-optic systems. In this mode (sometimes referred to as *end-light*), the end of the fiber is exposed and delivers all the available light. Lateral mode (or *side-light*) is the complement to this and is often used to simulate neon. Lightly Expressed does not supply lateral-mode fiber-optic products.

ballast Device used with discharge lamps for stabilizing the current in the discharge.

baseboard A type of heating distribution equipment in which either electric resistance coils or finned tubes carrying steam or hot water are mounted behind shallow panels along baseboards. Baseboards rely on passive convection to distribute heated air in the space. Electric baseboards are an example of an individual space heater.

beam axis The direction in the center of the solid angle which is bounded by directions having luminous intensities of 90 percent of the maximum intensity of a luminaire.

beam efficiency The ratio of the flux emitted within the solid angle defined by the beam spread, to the bare lamp flux.

beam lumens of a projector The quantity of light contained in that part of the beam at which $I = \frac{1}{2} I_{max}$ (Europe) or $I = \frac{1}{10} I_{max}$ (United States).

beam spread The angle (in the plane through the beam axis) over which the luminous intensity drops to a stated percentage of its peak intensity.

bipole speaker One type of surround speaker. In this instance two or more drivers are facing different directions, and their cones vibrate in phase. This causes an omnidirectional sound.

blended-light lamp Lamp containing in the same bulb a high-pressure mercury vapor discharge tube and an incandescent lamp filament connected in series. The bulb may be diffusing or coated with a fluorescent material, e.g., the MLL lamp.

BNC connector A connector commonly found with coaxial cable.

boiler A type of space-heating equipment consisting of a vessel or tank in which heat produced from the combustion of such fuels as natural gas, fuel oil, or coal is used to generate hot water or steam. Many buildings have their own boilers, while other buildings have steam or hot water piped in from a central plant. For this survey, only boilers inside the building (or serving only that particular building) are counted as part of the building's heating system. Steam or hot water piped into a building from a central plant is considered district heat.

brightness (The term *luminosity* is obsolete.) Attribute of visual sensation according to which an area appears to emit more or less light. *Note:* Brightness, according to the definition, is also an attribute of color. In British recommendations, the term *brightness* is now reserved to describe brightness of color; luminosity should be used in all other instances.

building energy manager A person whose chief day-to-day responsibility is the physical operation and maintenance of the building's heating and/or cooling equipment.

building shell (envelope) The thermal envelope of the building, i.e., the roof, exterior walls, and bottom floors that enclose conditioned space through which thermal energy may be transferred to or from the exterior.

building shell conservation features Features designed to reduce the energy loss or gain through the shell or envelope of the building. This category includes roof, ceiling, or wall insulation; storm windows or double- or triple-pane glass (multiple glazing); tinted or reflecting glass or shading films; and exterior or interior shadings or awnings.

channel leakage Leakage occurs with matrix-surround encoded material. Sound meant to be heard from one channel is also heard from another channel. The solution comes with the new 5.1-channel Dolby Digital and six-channel DTS sound systems by virtue of a discrete channel sound system.

codec (coder/decoder) Any hardware device or algorithm that converts analog video or audio between uncompressed analog and compressed digital formats.

collision The loss of electronic signals, or packets, that results when two workstations attempt to transmit data simultaneously across a shared medium. Data must be resent as a consequence.

color rendering General expression for the effect of an illuminant on the color appearance of objects in conscious or subconscious comparison with their color appearance under a reference illuminant.

color rendering index (CRI) Measure of the degree to which the psychophysical colors of objects illuminated by the source conform to those of the same objects illuminated by a reference illuminant for specified conditions.

compact fluorescent lightbulb A lightbulb designed to replace screw-in incandescent lightbulbs; often found in table lamps, wall sconces, and hall and ceiling fixtures of commercial buildings with residence-type lights. They combine the efficiency of fluorescent lighting with the convenience of standard incandescent bulbs. Light is produced in the same way as with other fluorescent lamps. Compact fluorescent bulbs have either electronic or magnetic ballast.

contrast Subjective assessment of the difference in appearance of two parts of a field of view seen simultaneously or successively.

cornice lighting Lighting system comprising light sources shielded by a panel parallel to the wall and attached to the ceiling, and distributing light over the wall.

cove lighting Lighting system comprising light sources shielded by a ledge or recess, and distributing light over the ceiling and upper wall.

crossover cable A type of networking cable in which some wires are reversed from one end to the other to join two computers or two hubs.

CRT projector One type of front projector. It consists of three tubes, each putting out one color: red, green, and blue. They offer brightness and detail, but are difficult to set up, and convergence is required about twice a year. (See **front projector**.)

demand-side management (DSM) The planning, implementation, and monitoring of utility activities designed to influence customer use of electricity in ways that will produce desired changes in a utility's load shape (i.e., changes in the time pattern and magnitude of a utility's load). Utility programs falling under the umbrella of DSM include: load management, customer generation, and innovative rates. DSM includes only those activities that involve a deliberate intervention by the utility to alter the load shape. These changes must produce benefits to both the utility and its customers.

diffuse reflection Diffusion by reflection in which, on the macroscopic scale, there is no regular reflection.

diffused lighting Lighting in which the light on the working plane or on an object is not incident predominantly from a particular direction.

diffuser Device used to alter the spatial distribution of radiation and depending essentially on the phenomenon of diffusion.

digital light processor (DLP) Processor used to control digital micromirror devices (DMDs) in order to make extremely bright, sharp pictures. It may control one or three DMDs. (See **front projector**.)

digital micromirror device (DMD) A mirror that is very small (micromirror) that can be kept as is or tilted a certain number of degrees in order to reflect light. As such, it is either on or off. It can be turned on and off at various rates per second to achieve different levels of brightness. They are commonly used together to form micromirror *wafers* and are controlled by a digital light processor. (See **front projector**.)

digital versatile disc (DVD) Previously known as digital video disc, this represents the latest in home theater. It is a purely digital format using MPEG-1 and/or MPEG-2 compression. This may result in artifacts such as pixellation. The format also has the ability to have multiple aspect ratios, several different versions of a movie with several different captions, as well as Dolby Digital sound. Each disc consists of two layers so that when the end of one layer is reached, the laser beam focuses down to the next layer for a seamless layer change. Be sure to see the article on DVDs for more information.

dimmer A device in the electric circuit for varying the luminous flux from lamps in a lighting installation.

dipole speaker One type of surround speaker. In this instance two or more drivers are facing in different directions (most commonly and by definition 180°), and their cones are vibrating out of phase. This causes nulling out of the sound by the viewing area which forms a "figure-eight" sound field.

direct flux On a surface, the luminous flux received by the surface directly from the luminaires of the installation.

direct lighting Lighting by means of luminaires with a light distribution such that 90 to 100 percent of the emitted luminous flux reaches the working plane directly, assuming that this plane is unbounded.

directional lighting Lighting in which the light on the working plane or on an object is incident predominantly from a particular direction.

document-conferencing (data-conferencing) application Software that lets users on different machines share applications and jointly edit text and graphics files.

Dolby AC-3 The old name for the most popular 5.1-channel home theater sound system. Now called Dolby Digital, it consists of front left and right speakers, a center speaker, left and right surrounds, and a low-frequency effects (LFE) channel, usually used with a subwoofer.

Dolby Digital The new name for the most popular 5.1-channel home theater sound system. Formerly called Dolby AC-3, it consists of front left and right speakers, a center speaker, left and right surrounds, and a low-frequency effects (LFE) channel, usually used with a subwoofer.

Dolby Pro-Logic Most popular surround format. Almost any receiver nowadays has it. It uses matrixed surround in order to encode four channels of sound: left and right front channels, a center channel, and one surround channel. It is quite common to see two speakers used for the one surround channel, however, as well as a subwoofer to supplement the speakers.

Dolby Surround The encoding process used to make material compatible with Dolby Pro-Logic.

downlight Small luminaire concentrating the light, usually recessed in the ceiling.

DTS Coherent Acoustics A 5.1-channel sound format for use in home theaters. The channels are left and right front channels, left and right surround channels, a center channel, and a low-frequency effects (LFE) channel, usually used with a subwoofer.

duct A type of heating and/or cooling distribution equipment that is a passageway made of sheet metal or other suitable material to convey air from the heating, ventilating, and cooling system to and from the point of utilization. (See **air-handling unit**.)

economizer cycle A heating, ventilation, and cooling (HVAC) conservation feature consisting of indoor and outdoor temperature and humidity sensors, dampers, motors, and motor controls for the ventilation system to reduce the air-conditioning load. Whenever the temperature and humidity of the outdoor air are more favorable (lower heat content) than the temperature and humidity of the return air, more outdoor air is brought into the building.

energy-efficient ballast A lighting conservation feature consisting of an energy-efficient version of a conventional electromagnetic ballast. The ballast is the transformer for fluorescent and high-intensity discharge (HID) lamps and provides the necessary current, voltage, and waveform conditions to operate the lamp. An energy-efficient ballast requires lower power input than a conventional ballast to operate HID and fluorescent lamps.

energy management and control system (EMCS) An energy management feature that uses mini- and microcomputers, instrumentation, control equipment, and software to manage a building's use of energy for heating, ventilation, air conditioning, lighting, and/or business-related processes. These systems can also manage fire control, safety, and security. Not included as an EMCS are time-clock thermostats.

escape lighting That part of emergency lighting provided to ensure that an escape route can be effectively identified and used in case of failure of the normal lighting system.

Ethernet A LAN cable-and-access protocol that uses twisted-pair or coaxial cables and CSMA/CD (Carrier Sense Multiple Access with Collision Detection), a method for sharing devices over a common medium. *Ethernet* runs at 10 Mbits/s; *Fast Ethernet* runs at 100 Mbits/s.

FCIF (full common intermediate format) A 352-by-288 video format that is described by the ITU's H.261 specification. FCIF is sometimes called *CIF*.

feeder See *tail.*

fiber bundle The collection of individual fibers that supply light to the fixture. These fibers are held together and protected by the sheathing.

FiberOptic Variously *fiber-optic*, *fibre-optic* (England), and *fiberoptic*; refers to the conduction of light waves through materials of exceptional clarity and across long distances. Fiber optics demonstrate total internal reflection by combining like materials of differing indices of refraction.

floodlight Projector designed for floodlighting, usually capable of being pointed in any direction and of weatherproof construction.

fluorescent lightbulb This is usually a long, narrow white tube made of glass, coated on the inside with fluorescent material, that is connected to an electric fixture at both ends of the lightbulb. The tube may also be circular or U-shaped. The lightbulb produces light by passing electricity through mercury vapor, causing the fluorescent coating to glow or fluoresce. Excluded are compact fluorescent lightbulbs, which are listed in a separate category. Fluorescent lightbulbs are included in standard fluorescent in the lighting equipment category.

fractional T1 A digital phone service that provides a portion of a T1 line's full 1.544-Mbits/s bandwidth. Fractional T1 lines are usually partitioned in 56 kbits/s increments and are sometimes used to provide 384 kbits/s service for high-quality videoconferencing applications.

frame rate The number of images per second displayed in a video stream. Approximately 24 frames per second (fps) is considered full-motion video.

front projector One type of viewing device. This is a separate unit that projects the image onto a separate screen, allowing screen sizes of over 300 in.

full-duplex Use of simultaneous two-way communication between network cards, which effectively doubles the available bandwidth from 10 to 20 Mbits/s in Ethernet or from 100 to 200 Mbits/s in Fast Ethernet.

G.711, G.722, G.728 H.320's three audio compression standards. *G.711* defines the common A-law and mu-law codecs and specifies a 64 kbits/s 3.4-kHz compressed-audio stream. *G.722* describes higher-quality 64-kbits/s 7.0-kHz audio, and *G.728* specifies a more efficient 16-kbits/s 3.4-kHz audio stream.

general lighting Substantially uniform lighting of an area without provision for special local requirements.

glare Condition of vision in which there is discomfort or a reduction in the ability to see significant objects, or both, due to an unsuitable distribution or range of luminance or to extreme contrasts in space or time.

glass fiber The original fiber-optic material, still the standard in communications technology. Glass fiber requires a large bend radius and is not easily field-terminated. Lightly Expressed does not typically use glass fiber.

group videoconferencing system: A relatively large videoconferencing system that links groups of people located at remote sites.

H.261 The video codec component of H.320.

H.310 A suite of H.320-like recommendations that define videoconferencing over broadband ISDN and ATM and use MPEG-2 compression.

H.320 A suite of ITU (International Telecommunications Union) recommendations that define videoconferencing mechanisms over switched digital services such as ISDN, fractional T1, and switched 56. H.320 incorporates standards such as H.261, G.711, G.722, and G.728.

H.321 A suite of H.320-like recommendations that define videoconferencing over high-speed, wide-area networks such as broadband ISDN and asynchronous transfer mode. H.321 uses H.261 compression and is backward-compatible with the H.320 standard.

H.322 A suite of H.320-like recommendations that define videoconferencing over isochronous networks such as IEEE 802.9a IsoEthernet.

H.323 A suite of H.320-like recommendations that define videoconferencing over packet-switched local-area networks such as Ethernet and Token-Ring.

H.324 A suite of H.320-like recommendations that define videoconferencing over POTS phone lines.

halogen lamp Gas-filled lamp containing a tungsten filament and a small proportion of halogens.

halogen lightbulb A type of incandescent lightbulb that lasts much longer and is more efficient than the common incandescent lightbulb. The lightbulb uses a halogen gas, usually iodine or bromine, that causes the evaporating tungsten to be redeposited on the filament, thus prolonging its life.

hard-matte A filming technique in which plates block out the top and bottom of the picture as it is being filmed in order to achieve a wide-screen effect. The opposite is soft-matte.

high-definition television (HDTV) The supposed "future" of television. It is a viewing format with a supposed aspect ratio of 16:9/1.78:1. It is slated to have more than 1000 lines of resolution, as well as to have Dolby Digital be the official sound format. There are two forms in existence. There is an analog system in Japan, and in the United States a digital system is proposed by the Grand Alliance. This system is supposed to coexist with and eventually replace NTSC around the year 2006.

high-intensity discharge (HID) lightbulb A lamp bulb that produces light by passing electricity through gas, which causes the gas to glow. Examples of HID lamps are mercury vapor lamps, metal halide lamps, and high- and low-pressure sodium lamps. HID lamps have an extremely long life and emit many more lumens per fixture than do fluorescent lights.

high-pressure sodium (vapor) lamp Sodium vapor lamp in which the partial pressure of the vapor during operation is on the order of 104 Pa, for example, SON and SON-T lamps.

hub A hardware device that serves as the junction at which individual PCs and other network devices connect to one another. Up to four hubs can be connected per LAN segment to increase the number of available ports.

illuminators Source of light for fiber-optic lighting. The illuminator consists of a transformer, ballast, lamp holder, lamp, and fan. There are many different illuminator manufacturers, as well as different lamp intensities, color temperatures, and lamp types. Illuminators may come equipped with color-change wheels, dimmers (mechanical or electrical), and remote control capability.

incandescent lightbulb A lightbulb that produces a soft warm light by electrically heating a tungsten filament so that it glows. Because so much of the energy is lost as heat, these are highly inefficient sources of light.

increased safety luminaire: Enclosed luminaire that satisfies the appropriate regulations for use in situations where there is risk of explosion.

indirect lighting Lighting by means of luminaires with a light distribution such that not more than 10 percent of the emitted luminous flux reaches the working plane directly, assuming that this plane is unbounded.

infrared radiation Optical radiation for which the wavelengths are longer than those for visible radiation.

ISDN (Integrated Services Digital Network) A type of digital telephone service available in two speeds: 128-kbits/s basic-rate interface (BRI) and 1.54-Mbits/s primary-rate interface (PRI).

iso-intensity curve Curve traced on an imaginary sphere with the source at its center and joining all the points corresponding to those directions in which the luminous intensity is the same, or a plane projection of this curve.

LAN Local-area network; a communications network that links PCs and other devices in a single office or small campus. In *client/server LANs*, users' PCs (the clients) access shared files and sometimes applications stored on a dedicated PC that acts as the server. In *peer-to-peer networks*, any connected PC can serve as both a client and a server.

laserdisc Looking like an oversized CD, currently one of the best viewing media for home theater. It has a theoretical resolution of 425 lines. It also has four audio channels: two analog and two pulse code modulation (PCM) digital tracks.

lateral mode The least efficient mode of light conduction in a fiber-optic system. In this mode, light is emitted from the sides of the fiber-optic strand. As a result, there is significant light loss (this translates to uneven linear light distribution) along the length of the fiber. Axial mode utilizes fiber-optic systems in their most efficient manner (exposing the end of the fiber and delivering all available light) and, as a result, saves on material and energy costs.

LCD projector One type of front projector. It is the smallest type of all. It is sort of like a Watchman with a magnifying glass, but more detailed. One major benefit is that convergence is not required. One drawback is that this technology results in pixellation. (See **front projector**.)

leakage See **channel leakage**.

letterbox This process, which is used on many laserdiscs and some TV broadcasts, is used to achieve a wide-screen image, where the image is considerably wider than standard NTSC fare. The end result is a wider picture with black bands on the top and bottom of the screen, which reduces the overall resolution of the image. Another option with greater detail, but one that is less widely used, is to anamorphically squeeze the picture.

light-valve projector One type of front projector. It combines the technologies of LCD projectors and CRT projectors. It offers exceptional detail and brightness.

line doubler, tripler, or quadrupler Device that doubles, triples, or quadruples the number of lines that make up a picture, therefore increasing detail and ridding the picture of scan lines. It is usually used with front projectors.

local lighting Lighting for a specific visual task, in addition to and controlled separately from the general lighting.

localized lighting Lighting designed to illuminate an area with a higher illuminance at certain specified positions, e.g., those at which work is carried out.

luminaire Apparatus that distributes, filters, or transforms the light given by a lamp or lamps and which includes all the items necessary for fixing and protecting these lamps and for connecting them to the supply circuit. Luminaires are classified as having narrow, average, or broad spread.

manual dimmer switch A lighting conservation feature that changes the level of light in the building. These are like residential-style dimmer switches, which are not commonly used with fluorescent or HID lamps.

matrixed surround Term used to describe the process to make Dolby Pro-Logic compatible material. It fits four channels of sound into a space meant for two channels. The center channel is decoded by using material common to both left and right channels, and the surround channel is decoded by extracting the sounds with inverse waveforms. This process results in channel leakage.

mercury vapor lamp An electric discharge lamp, with or without a coating of phosphor, in which during operation the partial pressure of the vapor is on the order of 105 Pa, for example, HPL and HPL-N lamps.

metal halide A very bright, high-efficiency lamp with a high color temperature and a long lamp life.

monopole speakers One type of speaker with all drivers facing one direction. Used for precise placement of sounds. Usually used in front and center speakers.

multipoint control unit (MCU) A device that links three or more point-to-point videoconferencing systems into a multipoint conference.

natural lighting control sensors A lighting conservation feature that takes advantage of sunlight to cut the amount of electric lighting used in a building by varying output of the lighting system in response to variations in available daylight. They are sometimes referred to as *daylighting controls* or *photocells.*

NDIS Network Driver Interface Specification; a hardware- and protocol-independent driver specification for network adapter cards developed by Microsoft and 3Com and supported by many NIC vendors and operating systems.

NIC Network interface card; another name for the adapter card that goes into your PC and connects it to a LAN. *Note:* The term is often applied to lighting designed to illuminate a particularly small area, e.g., a desktop.

NTSC The standard by which TV is broadcast in the United States. It has a theoretical maximum resolution of 525 lines and an aspect ratio of 4:3 or 1.33:1. (See **common aspect ratios**.)

occupancy sensor A lighting conservation feature that uses motion or sound to switch lights on or off; also known as *ultrasonic switching.* When movement is detected, the lights turn on and remain on as long as there is movement in the room. Occupancy sensors that detect sound work similarly to ultrasonic switching; when sound is detected, the lights turn on. In this report, occupancy sensors refer to detecting movement, not sound.

ODI Novell's Open Data-link Interface, a device driver specification for network adapter cards. Like NDIS, ODI is protocol-independent and is supported by many NIC vendors and operating systems.

packet An individual bundle of data transmitted across the network. Each packet includes information about its size, origin, and destination as well as error detection and correction bits, in addition to the data being sent.

PAL The standard by which TV is broadcast in Europe. It has a theoretical maximum resolution of 625 lines and an aspect ratio of 4:3/1.33:1, and in some places 16:9/1.78:1.

pan and scan A technique in which the right and/or left edges of widescreen material are chopped off in order to fit the picture into a narrower aspect ratio. People who do this select the best part of the image to scan, and then if the whole image needs to be seen, they scan across the rest of the frame. (See **common aspect ratios**.)

personal computer (PC) A self-contained electronic system with all the components necessary to perform computerized functions including a screen (monitor), keyboard and/or mouse, and central processing unit.

personal videoconferencing system A system designed to connect a single person into a teleconference. Most desktop videoconferencing systems are personal systems.

photovoltaic (PV) arrays: A renewable-energy feature that is a device that produces electric current by converting light or similar radiation.

POTS (plain old telephone service) Conventional analog telephone service.

QCIF (Quarter Common Intermediate Format) A 176-by-144 video format defined by the ITU's H.261 specification.

quartz halogen Lamps that burn hotter and tend to have a shorter lamp life than metal halide lamps. They do demonstrate, however, a lower color temperature which is desirable in many applications.

rear projector One type of viewing device. It is essentially a front projector and screen rolled into one. Screen sizes range from 45 to 80 in.

rear speaker The term wrongly applied to surround speakers.

receiver The guts of many home theaters. It has a decoder, audio/video switcher, AM/FM tuner, and amplifier section all in one.

recessed luminaire Luminaire mounted above the ceiling or behind a wall or other surface so that any visible projection is insignificant.

reflector Device in which the phenomenon of reflection is used to alter the spatial distribution of the luminous flux from a source.

reflector lamp Lamp in which part of the bulb is coated with a reflecting material, either diffuse or specular, so as to control the light, for example, HPL-R, MLR, and 'TL'F lamps.

refresh rate The rate at which the picture redraws itself in 1 second. Usually expressed in hertz (Hz).

resolution A term associated with the number of lines that make up the vertical portion of the picture. The higher the number, the more detailed the picture is. If the resolution is too low and the picture size too big, you get scan lines.

RJ-45 port The standard eight-pin connector found in hubs and NICs. Twisted-pair cable runs between two RJ-45 ports.

scalloping An effect of unblended areas of light when the light distribution points are too close to a reflective surface.

screen What the picture is projected onto. The screen is more important when it comes to front projectors, because the screen must be bought separately.

sheathing Plastic tube that protects the fiber bundle. Sheathing is available in many different materials and will be specified depending on your application.

soft-matte A projection technique in which plates block out the top and bottom of the picture as it is being projected in order to achieve a wide-screen effect. The opposite is hard-matte.

Sony Dynamic Digital Sound (SDDS) An eight-channel sound format used in commercial movie theaters. The eight channels are left front, left/center front, center front, right/center front, right front, left surround, subwoofer, and right surround. The sound is encoded between the sprockets on the film. No follow-up has been announced for home theater.

spotlight A (small) projector giving concentrated light of usually not more than 20° divergence.

spread Quantity of a luminaire that indicates the extent to which the light is spread out across the road.

standard fluorescent lightbulb See **fluorescent lightbulb.**

subwoofer A separate speaker used to handle the bass of movie soundtracks. It can be used with the low-frequency effects channel in the new digital sound formats. These speakers can sometimes handle frequencies as low as 15 Hz.

surround sound The popular term used to describe an experience where the sound surrounds you. This is best achieved using surround-encoded material, a receiver, and surround speakers.

surround speaker Type of speaker that best achieves the surround sound effect. This type of speaker diffuses the sound so as to make it harder to discern where the sound is coming from. Two popular types are bipoles and dipoles.

switched 56 service A type of dial-up digital phone service that provides 56 kbits/s bandwidth.

T.120 An ITU recommendation that standardizes document conferencing over a variety of transmission media. T.120 support is required for data conferences using equipment from multiple vendors.

T1 A type of 1.54 Mbits/s digital phone service, sometimes used for extremely high-quality videoconferencing.

tail The sheathed fiber bundle that connects the illuminator to the fixture. Tail length is typically limited to 30 ft, as longer runs will begin to show lighting differentials in the fixture.

terminating Cutting the ends of the individual fibers all at once so that the finished surface of the bundle is uniform. To maximize the transmission of light between the illuminator and the end of the fiber bundle, the fibers must be properly terminated.

THX Rumored to stand for the Tomlinson Holman eXperience. Others say it is named after George Lucas' first film, "THX-1138." THX is a set of standards by which laserdiscs and videotapes are made, as well as by which home theater equipment is made. They are supposed to yield the highest quality in home theater.

tubular fluorescent lamp Gas-filled lamp containing halogens or halogen compounds, the filament being of tungsten.

ultraviolet radiation Optical radiation whose wavelengths are shorter than those for visible radiation.

uplink port A jack on a hub that connects slower devices and faster backbones, such as a 10 Mbits/s Ethernet LAN and a 100 Mbits/s Fast Ethernet backbone.

vapor-tight luminaire Luminaire so constructed that a specified vapor or gas cannot enter its enclosure.

variable air volume (VAV) system An HVAC conservation feature usually referred to as *VAV* that supplies varying quantities of conditional (heated or cooled) air to different parts of a building according to the heating and cooling needs of those specific areas.

water-heating equipment Automatically controlled, thermally insulated equipment designed for heating water at temperatures less than 180°F for other than space-heating purposes. This survey collected data to distinguish between two types of water-heating equipment: centralized and distributed.

whiteboard A document-conferencing function that lets multiple users simultaneously view and annotate a document with pens, highlighters, and drawing tools. More advanced whiteboard programs handle multipage documents and provide tools for delivering them as presentations.

wide-screen Term used to describe a picture in which the aspect ratio is wider than the NTSC standard of 4:3/1.33:1. Almost all movies made nowadays are shot in some wide-screen format. To solve the problem of different aspect ratios, several different techniques can be used. Among them are anamorphic squeezing, letterboxing, and panning and scanning.

work (or working) plane Reference surface defined as the plane at which work is usually done.

INDEX

AC isolation transformers, 176, 177
AC wiring, avoiding high-voltage, 59–60
Access entry devices, 88
Accessibility, 9
Acoustics, home theater, 150–154
Air circulation, 85–92
  attic ventilator for, 85
  whole-house fans for, 85–86
Air cleaners, 80–81
Alpan, 276
AMP, Inc., 17, 29, 58, 132, 133
Amplification, sound, 159
Anamorphic squeezing, 172
Anti-icing equipment, 266–267
Antisiphon valves, 254–257
Apartment (multiunit) housing, 12–13
Apprehension (with security systems), 219–220
Aspect ratios, 167, 171–173
AT&T Bell Laboratories, 136
Attenuation, 56, 184
Attic ventilator, 85
Automatic dialers, 216–219
Automatic drain valves (watering systems), 259
Automation closet, 62–64

Backflow preventer (watering systems), 257–259
Balanced differential mode, 140
Ballast (lighting), 226
Baseband video, 54
Bathrooms:
  controllers in, 98
  faucets, automatic, 99–102
  interfaces in, 98–99
  toilets, automatic, 101–103
Benefits of home automation:
  to builders/installers/service contractors, 10–12
  to homeowners, 5, 8–10
Brightness, task, 228
Brightness ratios, 228
Builders:
  benefits of home automation to, 10–12
  role of, 4–5
Bus topology, 124–126
BusCall, 145
Business phone lines, 120

Cable television (CATV), 43
Cables/cabling (*see* Fiber optics; Wires/wiring)
CAL (Common Application Language), 23
Carbon monoxide detectors, 213–215
Card readers, 195–196
Cathode-ray tubes (CRTs), 163–164
CATV (cable television), 43
CCTV (*see* Closed-circuit television)
CEBus (*see* Consumer Electronic Bus)
CEBus Industry Council (CIC), 24
Ceiling-mounted sensors, 246
Ceiling-mounted speakers, 64–65
Ceilings:
  and acoustics, 152–153
CEMA (*see* Consumer Electronics Manufacturers Association)
Checklist, planning, 6–8
CIC (CEBus Industry Council), 24
Clock switches, 246
Closed-circuit television (CCTV), 8, 44, 197–198
Clothes dryers, 114–116

Clothes washers, 113–114
Coaxial cable, 53–58, 271–272
Combiners, 55, 183–184
Command pathways, 43–67
  and automation closet, 62–64
  coaxial cable, 53–58
  ducts, wiring, 66
  empty conduits, installation of, 65
  fiber optics, 58–59, 66–67
  fishing wires, 67
  low-voltage wiring, 48–53
  power line wire, 44–49
  raceways, 65–66
  routing, wire, 59–62
  speakers, ceiling, 64–65
Common Application Language (CAL), 23
Communication(s), 119–146
  with builders, 5
  and convergence, 144–146
  dual coaxial cable wiring systems, 132–133
  fiber optic, 135–136
  networks, 133–135
  serial interfaces, 136–144
  telephone lines, 119–120, 127–132
    business phone lines, 120
    connecting blocks for, 129–132
    family phone lines, 120
    polarity, 128–129
    running, 131–132
    surface-mounted screw terminals, 129–130
  topologies, 120–128
    bus, 124–126
    hybrid, 126–128
    ring, 123–124
    star, 121–123
Complexes, 106–108
Composite video, 181
Compression, data, 268
Computer controllers, 3
Connecting blocks, 129–132
Consumer Electronic Bus (CEBus), 22–24, 31–34, 55, 133, 244
Consumer Electronics Manufacturers Association (CEMA), 22, 27–28
Contractors, benefits of home automation to, 10–12
Contrast ratios, 228
Convenience of home automation, 8
Convergence, 144–146, 267–270
Conversion tables, 293–296
Cooperative Research and Development Act of 1984, 28–29
Couplers, power line, 14
C-Phone Home, 273
CRTs (*see* Cathode-ray tubes)
Custom Solutions, Inc., 274
Customer satisfaction, 11

Daisy chaining, 49
Dedicated controllers, 37–38
Denon, 275
Deregulation, utility, 145
Dialers, automatic, 216–219
Diffusion, sound, 152
Digital satellite systems (DSSs), 55, 177–181
  equipment for, 177–178
  formats, 179
  installation of, 179–181
  providers of, 179
Digital surround (DTS), 156, 159
Digital versatile discs (DVDs), 176, 177, 275
Direct sound, 150
Disney Studios, 154
Dolby system, 154–156, 177
Doors:
  exterior, 200–201
  garage, 201–205
  main entry, 90–92
  patio, 90
  and security, 198, 200–205
Drops, 56
Dryers, clothes, 114–116
DSSs (*see* Digital satellite systems)
DTS (*see* Digital surround)
Dual coaxial cable, 57–58, 132–133
Duct boosters, 83–85
Duct dampers, electronic, 82–83
Ducts, wiring, 66
DVDs (*see* Digital versatile discs)

EIA (*see* Electronics Industry Association)
EIA-708 standard, 27–28

Electric solenoid valves, 95–96
Electronic duct dampers, 82–83
Electronics Industry Association (EIA), 22, 24, 27, 137
Empty wire conduits, 65
End-of-line (EOL) resistors, 217–219
Energy conservation:
  as benefit of home automation, 5
  and lighting, 241
Energy recovery ventilators (ERVs), 77
Energy-saving (ES) lamps, 232–237
Entertainment (*see* Home theater)
EOL resistors (*see* End-of-line resistors)
Equal-potential grounding, 45
ERVs (energy recovery ventilators), 77
ES lamps (*see* Energy-saving lamps)
Ethernet, 134–135
Exterior doors, 200–201
Exterior lighting, 242–243

Family phone lines, 120
Fans, whole-house, 85–86
Fantasia, 154
Faraday cage method, 45
Faucets, automatic, 99–102
Federal Communications Commission (FCC), 37, 49, 53
Fences, 195
Fenestration, 227
Fiber optics, 271
  as command pathway, 58–59
  for communications, 135–136
  enclosure systems for, 66–67
  and lighting, 241–242
File server, 134
Filters, power line, 14
Finishes:
  room, 227
  task, 227
Fire security, 205–213, 215–216
FireWire, 141–144
First reflections, 151
Fishing wires, 67
Fixtures, lighting, 237–242
Flat cathode-ray TV tubes, 165–166
Floors, 152
Fluorescent lamps, 230–232
Franklin cone, 45
Front-projected curved screens, 168–169
Front-projected flat screens, 169–170
Front-projection two-piece television, 164

Gain, projection screen, 167–168
Garage doors, 201–205
Gas plumbing, 95
Gates, 88, 195–197
GFCIs (ground-fault circuit interrupters), 97
Glass break detectors, 196–197
Global positioning system (GPS), 145
Graphical-user interface (GUI), 38
Graphics projectors, 166–167
Greyfox Systems, 58, 132, 133
Ground-fault circuit interrupters (GFCIs), 97
Grounding, 45
GUI (graphical-user interface), 38

Halogen lamps, 232–236
HDTV (*see* High-definition television)
Heat detectors, 208–213
  circuits for, 209–211
  connections for, 213
  fitting/fastening, 211–212
  mechanical protection of, 212–213
Heating, ventilation, and air conditioning (HVAC), 71–92
  and air circulation, 85–92
  ducts, 82–85
  humidistats, 80
  thermostats, 73–80
  zoned control, 71–73
High-definition television (HDTV), 167, 173
Home Director, 274
Home offices, 11
Home Plug and Play (HomePnP), 24
Home run (*see* Star topology)
Home theater, 149–189
  and acoustics, 150–154
  audio components, 154–162
  digital satellite systems, 177–181
  DVD, 177
  interfacing with, 187–189
  laserdisc, 176–177
  Lucasfilm THX, 156–157
  and noise interference, 173–177

Home theater (*Cont.*):
  video components, 159, 162–173, 181–187
HomeVision-PC, 274
Honeywell, 31
House codes (X-10), 18
Humidistats, 80
100baseT format, 135
HVAC (*see* Heating, ventilation, and air conditioning)
Hybrid topologies, 126–128

IBM, 274
IDCs (insulation displacement connectors), 130
IEEE 1394 interface, 141–144
IEEE (Institute of Electrical and Electronics Engineers), 141
IES Technologies, 58, 132, 133
Impact tool and blade, 130
Incandescent lamps, 229
Inferior state, 23
Infrared (IR) control systems, 32, 34–36
  coaxial repeaters, 36
  low-voltage repeaters, 34, 36
  RF repeaters, 36
Infrared (IR) reflecting film lamps, 233–235
Infrared remotes, 196
Installers, benefits of home automation to, 10–12
Institute of Electrical and Electronics Engineers (IEEE), 141
Insulation displacement connectors (IDCs), 130
Intelligent home products, 4
Interactive TV (ITV), 185–187, 267
Interfaces:
  in bathrooms/spas, 98–99
  graphical-user interface, 38
  home theater, 187–189
  with PC controllers, 38–40
  serial, 136–144
Internet, 9, 10
Interoperable systems, 21
IR (*see* Infrared *entries*)
ITV (*see* Interactive TV)

Joists, drilling holes in, 61

Keypads, 196
Keyswitches, 196

Lamps, 226
LANs (*see* Local-area networks)
Laserdiscs (LDs), 176–177
Laundry areas, 112–116
  dryers, 114–116
  washers, 113–114
LCD projectors, 164
LDs (*see* Laserdiscs)
Leak sensors, 96
Letter boxing, 172
Levels of interaction, 4
Leviton, 274
LIA (*see* LonMark Interoperability Association)
Lighting, 3–4, 225–246
  components of, 226–228
  controls for, 243–246
  exterior, 242–243
  fixtures, 237–242
    recessed lighting, 237–238
    remote-source lighting, 239–241
    track lighting, 238, 239
    wall sconces, 238, 240
  for pools/spas, 111–112
  and security, 194, 195
  types of, 229–237
    energy-saving, 232–237
    fluorescent, 230–232
    incandescent, 229
Lightning rods, 45
Line doublers, 181–182
LM334 temperature sensor, 77, 78
LM335 temperature sensor, 77–79
Local-area networks (LANs), 133–135
Locks, 195
LonMark Interoperability Association (LIA), 25–26
LonWorks, 24–28, 32–34, 244
Loudspeakers:
  ceiling-mounted, 64–65
  home-theater, 157–162
Low-voltage wiring, 48–53
  category-verified wiring, 50–52
  installation of, 52–53
Lucasfilm THX, 156–157

Lucent, 17
Luminaire, 226–227

M block, 130, 131
Macros, 188
Magnetic fluid conditioners, 108–109
Magnetohydrodynamics, 108
Manufacturers:
  of data communications cable, 281–285
  of security components/systems, 287–291
Market segments, 12
MediaCom, 144–145
Metal-oxide varistors (MOVs), 47
Modulators, 56–57
Molex, 29, 58, 132, 133
MOS (*see* Multimedia outlet system)
Motion sensors, 194–195
Motorized perimeter fence gates, 196, 197
MOVs (metal-oxide varistors), 47
Multimedia outlet system (MOS), 274–275
Multiservice central office switch, 10
Multiunit (apartment) housing, 12–13

NAD, 275
National Association of Home Builders (NAHB), 29
National Bureau of Standards (NBS), 207
National Electrical Code (NEC), 173
Navigation systems, 272–273
NBS (National Bureau of Standards), 207
NEC (National Electrical Code), 173
NetChannel, 187
Network Integrator Program, 28
Network interface devices (NIDs), 121–122, 124, 127, 129–131, 216
Networks, 120, 133–135
NIDs (*see* Network interface devices)
Nodes (LonWorks), 26–27
Noise filters, 46
Noise interference, 173–177
NTSC video, 1781

Occupancy sensors, 245–246
Optical fiber (*see* Fiber optics)

Pagoda Garden Light, 276
PAR-36 lamps, 236–237
Passive infrared (PIR) detectors, 194–195, 245
Pathways (*see* Command pathways)
Patio doors, remote-controlled, 90
Perforated projection screens, 170–171
Perimeter security, 194–198
Personal computer (PC) controllers, 38–40
PhoneMiser, 144–145
Photosensors, 246
Physically challenged individuals, accessibility for, 9
Pico Electronics Ltd., 17–18
PIR detectors (*see* Passive infrared detectors)
Planning checklist, 6–8
Plasma screen displays, 164–165
PLCs (*see* power line carriers)
Plumbing, 95–116
  electric solenoid valves, 95–96
  gas vs. water, 95
  for laundry areas, 112–116
  leak sensors, 96
  magnetic fluid conditioners, 108–109
  pools/spas, 111–112
  in showers/baths, 96–103
    controllers, 98
    faucets, automatic, 99–102
    interfaces, 98–99
    toilets, automatic, 101–103
  temperature balancing, 104–107
  water heaters, 110–111
  and water softeners, 106–108
Polarity (of telephone lines), 128–129
Pools, 111–112, 265–266
Power line carriers (PLCs), 2, 46–49
Power line wire, 44–49
  carrier devices with, 46
  carrier products, 47–49
  grounding, 45
  lightning protection with, 45
  surge protection, 46–47
PowerFlash module, 113–116
Profitability of home automation, 11

Programmable timers, 2–3
Projection monitors, 159
Projection screens, 167–171
  front-projected curved, 168–169
  front-projected flat, 169–170
  gain of, 167–168
  perforated, 170–171
  sizing of, 171
  stretched flat, 170

QuickSilver Hydra, 274

Raceways, 65–66
Radio-frequency control systems, 36–37
Radio-frequency ground breakers, 175
Radio-frequency (RF) remotes, 196
RCR (room cavity ratio), 227
Rear-projection television, 163–164
Recessed lighting, 237–238
Remote controls, wireless, 3
Remote-source lighting, 239–241
Research and licensing agreements, 29
Response time (with security systems), 220–221
Return path, 268, 272
Reverberation, 151
RF (*see* Radio-frequency *entries*)
RG-6 (coaxial cable), 55
RG-59 (coaxial cable), 55
Ring (telephone wire polarity), 128
Ring topology, 123–124
RJ-45 jack, 133
Room boundaries, 227
Room cavity ratio (RCR), 227
Room finishes, 227
Routing, wire, 59–62
RS-232 interface, 137–140
RS-485 interface, 140–141

Satellite, 270–271
Scale, 108
Sconces, wall, 238, 240
Scope of home automation, 3–5
Security, 2, 8, 193–221
  and apprehension, 219–220
  automatic dialers, 216–219
  with automation controllers, 38
  carbon monoxide detectors, 213–215
  closed-circuit television, 197–198

Security (*Cont.*):
  deterrents, 194, 195
  doors, 198, 200–205
  fire/smoke detectors, 205–208, 215–216
  heat detectors, 208–213
  manufacturers (list), 287–291
  perimeter security, 194–198
  philosophies of, 193–194
  and response time, 220–221
  sound discriminators, 196–197
  windows, 198–200
Sensors:
  ceiling-/wall-mounted, 246
  infrared, 245
  motion, 194–195
  occupancy, 245–246
  ultrasonic, 245
Serial interfaces, 136–144
  IEEE 1394 interface, 141–144
  RS-232 interface, 137–140
  RS-485 interface, 140–141
Service center (Smart House), 30–31
Service contractors, benefits of home automation to, 10–12
Set points, thermostat, 75
Set-top boxes, 268, 269
Sharp, 165, 166
Single-pole double-throw (SPDT) relays, 189, 219
Sirens, self-contained, 215–216
Smart House, 28–31, 33, 58, 132
Smoke detectors, 205–208
Sonex, 153
Sony Corporation, 165–166
Sound discriminators, 196–197
Spas, 111–112, 265–266
SPDT relays (*see* Single-pole double-throw relays)
Speakercraft, 275–276
Speakers (*see* Loudspeakers)
Splitters, 55, 182–184
Spread spectrum modulations, 23
Sprinkler risers, 250, 254
Sprinkler systems, automatic, 260–262
Standards, control, 17–40
  CEBus, 22–24
  dedicated controllers, 37–38
  Home PnP, 24

Standards, control (*Cont.*):
  infrared systems, 32, 34–36
  LonWorks, 24–28
  PC controllers, 38–40
  radio frequency systems, 36–37
  Smart House, 28–31
  Total Home, 31
  X-10, 17–21
Star topology, 50, 121–123
Stretched flat projection screens, 170
Subcontractors, benefits of home automation to, 10–12
Subscriber management, 268–269
Subwoofers, 159, 161
Superior state, 23
Surface-mounted screw terminals, 129–130
Surge protectors, 13–14, 46–47
Systems integration, 4

Task brightness, 228
Task finishes, 227
Task size, 228
Telecommunications Industry Association (TIA), 51
Telephone lines, 119–120, 127–132
  business phone lines, 120
  connecting blocks for, 129–132
  family phone lines, 120
  polarity, 128–129
  running, 131–132
  surface-mounted screw terminals, 129–130
TeleSite USA, Inc., 274
Television:
  closed-circuit, 44, 197–198
  direct-view, 163
  flat cathode-ray TV tubes, 165–166
  front-projection two-piece, 164
  high-definition, 167, 173
  interactive, 185–187, 267
  rear-projection, 163–164
  sound quality of, 166
  wide-screen, 171, 173
Temperature balancing, 104–107
Temperature control, 9
10base2 format, 135
10base5 format, 135
10baseT format, 135
Thermostats, 73–80
  approaches to controlling, 73–74
  connection/biasing of, 78–80
  programmable, 74–75
  protocol with, 75–76
  true communicating, 76–78
Thinnet, 135
TIA (Telecommunications Industry Association), 51
Timer controllers, 2–3
Tip, 128
Toilets, automatic, 101–103
Topologies, 120–128
  bus, 124–126
  hybrid, 126–128
  ring, 123–124
  star, 121–123
Toshiba Corporation, 165
Total Home, 31
Track lighting, 238, 239
Transducers, 161–162
Transformers, AC isolation, 176, 177
Tungsten-halogen lamps, 232–236
Twisted pair, 271, 272
Type 66 blocks, 130–131

UART, 139
UL (Underwriters Laboratories), 52
Ultrasonic sensors, 245
Underwriters Laboratories (UL), 52
Uninterruptible power supply (UPS), 47
Unit codes (X-10), 18
Unity gain, 56, 184
Unshielded twisted-pair (UTP), 134
UPS (uninterruptible power supply), 47
U.S. Tec, 58, 132, 133
USART, 139
Utility deregulation, 145
UTP (unshielded twisted-pair), 134

VCRs (*see* Videocassette recorders)
Video:
  baseband, 54
  cable for, 53–58
  home theater, 159, 162–173, 181–187
  NTSC, 181
  transmission of, 270–272
Video ground breakers, 176
Video hum bars, 173, 174

Videocassette recorders (VCRs), 149–150
VocalNet, 273
Voice browsers, 9

Walls:
 and acoustics, 154
 sconces on, 238, 240
 sensors on, 246
WANs (wide-area networks), 133
Warble, 216
Washers, clothes, 113–114
Water clarity, 109
Water heaters, 110–112
Water softeners, 106–108
Watering systems:
 interior, 262–265
 outdoor, 249–262
  antisiphon valves, 254–257
  automatic drain valves, 259
  backflow preventer, 257–259
  flushing of, 259–260
  installation, 250, 260–262
  risers, 250, 254
  sprinkler systems, automatic, 260–262
  trench depth, 258–259
  wires/splices, 257–258
WavePlane Technology, 275–276
WebTV, 186–187, 268–269
Whole-house blockers, 46
Whole-house fans, 85–86
Wide-area networks (WANs), 133
Wide-screen TV, 171, 173
Window coverings, motorized, 89–90
Windows:
 and lighting, 227
 remote-controlled, 86–88
 and security, 198–200
Wireless remote controls, 3
Wires/wiring:
 coaxial, 53–58, 271–272
 dual coaxial, 57–58, 132–133
 fishing wire, 67
 low-voltage wiring, 48–53
 manufacturers of (list), 281–285
 for networks, 134
 for outdoor watering systems, 257–258
 power line wire, 44–49
 routing of, 59–62
 for security systems, 219
 Smart House cabling, 29–30
 twisted pair, 271, 272
 video wiring, 182–186
 (*See also* Telephone lines)
Wiring distribution panels, 13
Wiring ducts, 66

X-10 protocol, 17–21, 33, 73–77, 113–116, 243–244
XbaseY format, 135

Yelp tone, 216

Z-Man Audio Signal Enhancer, 276
Zoned heating and cooling, 71–73